FACEBOOK SOCIETY

FACEBOOK SOCIETY

LOSING OURSELVES IN
SHARING OURSELVES

ROBERTO SIMANOWSKI

Translated by

SUSAN H. GILLESPIE

Columbia University Press *New York*

Columbia University Press
Publishers Since 1893
New York   Chichester, West Sussex
cup.columbia.edu
Copyright © 2018 Columbia University Press
Originally published in German as *Facebook-Gesellschaft.* © MSB:
Matthes & Seitz Berlin Verlagsgesellschaft mbH 2016.
All rights reserved

Library of Congress Cataloging-in-Publication Data
Names: Simanowski, Roberto, author.
Title: Facebook society : losing ourselves in sharing ourselves /
Roberto Simanowski ; translated by Susan H. Gillespie.
Other titles: Facebook-Gesellschaft. English
Description: New York : Columbia University Press, [2018] |
Includes bibliographical references and index.
Identifiers: LCCN 2017058609 (print) | LCCN 2017061792 (ebook) |
ISBN 9780231544344 (e-book) | ISBN 9780231182720
(cloth : alk. paper)
Subjects: LCSH: Facebook (Firm) | Facebook
(Electronic resource)—Social aspects. | Social networks.
Classification: LCC HM743.F33 (ebook) | LCC HM743.F33 S56
2018 (print) | DDC 302.30285—dc23
LC record available at https://lccn.loc.gov/2017058609

Columbia University Press books are printed on permanent
and durable acid-free paper.
Printed in the United States of America

COVER DESIGN: PHILIP PASCUZZO

For Luciana, best friend on- and offline

CONTENTS

PREFACE

'm so sick and tired of it! The hoary old complaints about how spoiled the younger generation is today. The decline of culture. Moral decadence. The alienation of human beings. The technological takeover of communication. Narcissistic flaunting of the self. Restlessness combined with apathy. Arthritis of the thumb and neck. Yet, in many cases, it was the complainers who started all this!

What was it like, twenty or thirty years ago, when they were sitting in the streetcar or the waiting room, buried in their books, in another world, as if the one around them didn't exist? Those fanatical time managers who read even standing up. Sometimes even while they were walking. Like the student in a French film about university life who is walking down the street engrossed in a book and sinks up to his knees in a pond. That was amusing. For it inspired sympathy when students, in trying to understand the world, completely forgot the world around them.

In real life, all that readerly zeal felt more like a kind of pressure. As if it would be a waste of time to be looking around at one's surroundings in the bus station, or spending hours staring out the train window and chatting with the passenger in the next seat, or being open, at the café, to whatever might come next. They, then, were always holding their book between themselves and the world, always had something to do, as if they didn't have a minute to lose. And today those same individuals complain that young people don't see the world around them!

Yes, admittedly, it can be uncanny to see everyone around us lost in their devices. You could pick your nose, make hideous faces, or murder someone in full public view without anyone noticing. On the other hand, how lost one could feel, staring straight ahead on the bus. How depressing, listening to the idle talk of colleagues over lunch. How menacing, the silence of the married couple in the restaurant. Instead of curvature of the spine, we should be talking about the happy faces that the smartphone is producing every day, eyes that are shining all over the world and among all generations—the impatient preschooler on an iPad in the waiting room, the elation of the eight-year-old sending a selfie to his friend on his mother's cell phone, the happiness of the taxi driver getting a WhatsApp message in a traffic jam. Even the elderly wax enthusiastic when their grandchildren appear on Skype. And what is more beautiful than a young woman looking contentedly at her smartphone? The sex appeal of a person who knows what she wants and doesn't have a moment to lose? How much more compelling than a student up to his knees in a pond!

Why, then, the outcry, given all the success? Why precisely from the people who started it all in the first place? How small-minded, to insist that there be at least one book on the device, not just Facebook or Candy Crush. How dishonest, to allow excitement only when it comes from focused reading of "worthwhile" texts? In this way, the fear of technology is paired with cultural conceit—unconsciously coupled with the mad hope that after you die life will have lost all its attractiveness anyhow. And it is not even clear why the serious book should still be preferred to distraction. All the criticism of the way young people respond to new media is based on an understanding of the world that was already outdated decades ago. That still places its bets on critique and the future instead of praising the present as it exists. Let them complain all they want about spoiled young people and cultural decline; let them mutually confirm that their era was a better era; let them assure us as convincingly as they can that now everything is going to the dogs—I cannot and will not hear any more of it.

Meanwhile, loudspeaker announcements on escalators warn us not to look at our mobile phones. Meanwhile, the police post warning videos showing texting pedestrians being brutally struck down by automobiles.[1] Meanwhile, it is a daily occurrence: passersby who stare into their cell phone the way the student once stared at his book. Resistance—defiantly (head back, eyes rolled heavenward) blocking the path of pedestrians who think they can make it across crowded intersections using only "ambient attention"—seems helplessly aggressive. They simply avoid you, without looking up. The only possible way to comment on this is to turn it into an image of the future (the anti-isolation cooking pot with built-in smartphone cradle) or a kind of game: Who will be the first to find an unbroken row of ten hunched-over, phone-absorbed passengers in the subway car? A better plan is secretly (in the waiting room or the pub) to use the new technology for old-time purposes: with Kant, not Facebook, on the screen. The smartphone as Trojan horse for academics—the current era is undoubtedly a fascinating one.

What is it that bothers us (if it does bother us) when we see people all around us immersed in their devices? What do we lack when they ignore us? Are we disappointed that they so brutally avoid the encounter with us? Are we concerned that they are running away from themselves? Why do we think differently about the student in the pond than we do about the smartphoners in the street? Is the value we ascribe to the obsession with media dependent on the number of people who succumb to it? On the type of medium and its contents? On the social model of its marketers? Historians of media know that almost every new medium was met with skepticism and disapproval from the older generation. The complaint about the cultural decline of the young goes back before the Christian era. Yet everyone who has ever given a smartphone to a parent also knows how enthusiastically it is possible, even as a skeptic, to talk about the new technology. The situation is anything but clear.

Thus, when it comes to new media, the more educated among its detractors try to resist the impulse to condemn it too quickly. There are plenty of criticisms: capitalizing on emotions, commercializing communication, self-marketing and self-surveillance, schooling in narcissism and banality, time wasting. . . . It is not as if these criticisms were wrong, in principle. But we need to reflect on the arguments that support them. The reproach of time wasting, for example, only makes sense if there is a normative concept of time utilization, as there was at the height of the Enlightenment, when it was written: "Reading merely to pass the time is immoral, for every minute of our lives is filled up with duties that we may not neglect without besmirching ourselves."[2] What sorts of duties fill up our lives today? Is there still a sociopolitical goal for which we ought to be putting ourselves on the line every day and every hour? Does the Enlightenment's famous *Sapere aude!* still call us to continuous self-perfection? Is political and ideological communication really better than the banal and commercial kind?

The value ascribed to the cultural forms that come along with new media is inevitably politically determined, but it is also determined by a person's philosophy of history and—no less important—is generationally specific. Teenagers talk about Facebook differently than retirees, just as they think about consumer culture differently than critical theorists. When we look at the present, we find a decreasing willingness to feel unhappy under circumstances that, from the perspective of critical theory, are catastrophic: a culture of consumerism, loss of the private sphere, alienation of the subject, environmental destruction, social ills, world-political tensions. . . . The trend toward acquiescence receives philosophical support from calls for positive thinking, affirmative emotions, and a childlike embrace of the world. That embrace may generally occur without orgiastic excess, but even restrained pleasure is enough to neutralize the impulse to negate the status quo, as critical theory aims to do. Society is more popular than its critics like, and the new media are, too.[3]

One of the main points of contact for all the smartphone users who people our urban landscapes is Facebook. Here, too, the values assigned to such problematic areas as advertising, privacy, and banality are indicative of an individual's attitude toward society as a whole. Here, as well, a too hasty rejection only obstructs the view of deeper problems. Why, despite the weighty objections, does Facebook continue to attract ever more users? Because most humans can be seduced? Because the lock-in effect breaks down all resistance? The first response is arrogant; the second leads straight to the next question: How is Facebook achieving a critical mass, from which hardly anyone manages to escape? This book starts from the assumption that, more than a decade after the founding of Facebook, it is time to ask questions that are not satisfied by all the right answers.

For Charles Taylor, the philosopher and political scientist, a given culture, no matter how odd its views and practices may seem to observers who hold different values, is legitimated by the simple fact of its existence over a long period of time.[4] The cultural values represented by Facebook (self-representation, transparency, interaction) are relatively young, but they undoubtedly enjoy broad acceptance. It may be too early to accord Facebook the legitimacy of long duration, which in light of the rapidity of digital development would be a contradiction in terms. But it is surely also too late to dismiss Facebook as an error or a fraud. The question is not by what dishonest means, for which impure purposes Facebook persuades its users to publicize their private lives. The question is: In what does the charm of this disclosure consist? What is the cultural basis for the lock-in? Why do so many people, so hopefully, still become Facebookers?

The short answer is: Facebook is cool, and it's fun. Beyond this, one generally hears the following reasons: Facebook gives people the exciting feeling of being a public person, with a history, a series of photographs, an audience, and fan letters; Facebook allows individuals to look into the lives of others, as a kind of "television" (remote seeing), with figures from their own biography as the characters; "friending" oneself effectively makes it possible to find the inputs and discussions that are of interest to a person: gossip, news, tips about events, political activism, cultural critique, academic links; Facebook allows communication at an extremely low interaction cost; you send to *everyone* and you receive from *everyone*, without the bother of having to address and confirm the receipt of messages; Facebook makes it possible to engage with different groups, on different themes, and it also conveys the feeling of being part of a community.

All these reasons are correct, yet they remain superficial. To understand Facebook, it is necessary to look beyond Facebook. Beyond the obvious, we need to understand Facebook as the answer to a problem that perturbs the (post)modern subject more or less consciously. It must be understood as the symptom of a cultural evolution that should be thought through the lens of a philosophy of history and should not be too quickly reduced to scenarios of political oppression or economic exploitation. The political-economic consequences of the Facebook system lie deeper. For one thing, through the accumulation and analysis of personal data Facebook generates knowledge as a tool of domination; in this way, it advances the process of commercialization. For another thing, through its invitation to a kind of experience of the self that is reflexively impoverished, it produces the very subjects who are no longer dismayed by this process.

This locates it within the trend of affirmative social relations, which it simultaneously promotes. Facebook is as popular as it is because it makes it possible to love the society we have. The more weight Facebook assumes as a symptom and motor of cultural development, the more appropriate it seems to speak of a *Facebook society*: a society whose forms of communication and cultural techniques are significantly determined by the practices of self-representation and world perception on Facebook. This concept of society is not restricted to those individuals who are actual members of Facebook, just as one does not have to own an automobile to be part of a car culture and feel its impact every day. At the same time, the concept of Facebook society is only acceptable if Facebook is understood metonymically: as a placeholder for comparable social networks that use similar technical and social *dispositifs* to teach society a specific way of thinking, feeling, and acting.[5] The central characteristics of Facebook society, in an overarching sense that exceeds any specific Facebook community, are all outcomes of these social networks: the disappearance of the present and the loss of reflective perception of both the world and oneself. Since both phenomena are advanced above all by Facebook, whose corporate leadership is, moreover, explicit about its ambitions to change society, this network is the most appropriate candidate for both conceptualizing and investigating these phenomena in detail.[6]

There are many buzzwords, mostly English or American in origin, that can be used to convey the essence of Facebook: hyperattention, multitasking, transparency, Big Data, immanence, interaction, immediacy, sharing, tagging, ranking, quantification, update, refresh, selfie, like, crowd, now. An essential phenomenon of the Facebook society, implied in many of these buzzwords, is the lack of contemporaneousness (*Zeitgenossenschaft*). This lack of a common temporality results from the paradoxical relationship of Facebook society to the present: It destroys the present by making it permanent. This sounds self-contradictory and counterintuitive; after all, the sharing culture that is practiced on Facebook (and other social networks) creates a situation in which more and more people document and present one another with virtually everything they experience. But precisely this compulsion to communicate prevents them from actually experiencing the present. The more or less reflexive, more or less unreflecting documentation of the lived moment replaces its real experience. By archiving the present—to anticipate one of this book's theses—we simultaneously negate, ignore, annul it; basically, we fall out of time precisely because we are permanently capturing it. This may recall Hegel's concept of "sublation" (*Aufhebung*), but in the dialectic of negation, preservation, and valorization on Facebook the third element has gone missing. The present is not raised epistemologically to a higher level but reduced to a lower one. For it is only the *distanced* proximity of reflection that allows us to understand the present: its complexity, its potential, its dark sides, and the alternatives that it forecloses. True contemporaneousness, according to the Italian philosopher Georgio Agamben, is precisely not possible in the mode of absolute immediacy. But Facebook society is a society of immediacy, of impatience and immersion.[7]

Distance, reflection, and immunity to the blinding glare of the present are what social networks such as Facebook overcome and prevent. They accomplish this through their use of spontaneous, visual, and automatized communications and potentially through the introduction of immersive augmented-reality technology. The future, Mark Zuckerberg declared in July 2015, lies in the immediate sharing of experience: "We will have AR and other devices that we can wear almost all the time in order to improve our experience and our communication." What Zuckerberg means by improving experience and communication is the end of language as a means of communication, as it is already practiced to some degree when, for example, via Snapchat, experiences are no longer grasped in words but preserved as an image. Facebook, and with it the entire "affective-computing" industry, thus appears as the twenty-first-century response to the crisis of representation, a crisis that was growing ever more acute in the twentieth. A cure for the unreliability of language is now available in the form of nonverbal documentation. Since language is *the* medium of reflection through which we assume a distanced stance to the world—distance that allows us to cognize it—every attempt to move beyond language is also a loss of contemporaneousness. This loss is magnified by the advance of "mathematecized thinking," which, as a further form of linguistic silencing, operates in thrall to numbers and in the form of algorithms. It is part and parcel of cybernetic concepts of futurity that, while they point far beyond Facebook society, essentially build on the latter's characteristics, especially its increasing datafication and simultaneous devaluation of reflection.[8]

The collective self-experience of Facebook society takes place in the framework of social networks that increasingly transcend cultural memory or grand narratives. This frees them from the claims of the past and the future on the present. People get connected, without regard to any coincidence of weltanschauung or ideological commitment, in the ritual of the technical. This bracketing of the rational and ideological makes possible a semblance of community that transgresses old boundaries and that is celebrated in social theories of the intuitive and of "erotic logic."[9] But does this already ensure the capacity to tolerate, suffer, and respect the other when she gets closer to us than a status update by a Facebook friend? How well armed is the lightness of being offered by social networks, unburdened by any serious debate, against new prophets who have simple answers to complex questions and are never at loss for a "truth"? The aggression that is increasingly evident when differing positions finally meet up suggests that on this level, too, the operating mode of Facebook society fails to develop the kind of

contemporaneousness that enables us, in the era of globalization and mass immigration, to develop the capacities that make
for a tolerant community of difference.

Against the background of this methodological consideration and theoretical perspective, the present book devotes particular attention to psychological, narrative, and political issues
and outlooks. Three theses serve as guidelines: First, behind
the narcissism of restless Facebook users is the fear of their
own experience, which is delegated to the network community
through communication of the given moment. Second, Facebook
more or less automatically, before narrative reflection, generates
an episodic "autobiography" whose actual narrator is the network and its algorithms. Finally, within the framework of their
superficial communication, social networks do create a cosmopolitan society that transcends political and cultural differences,
but in the process they do not develop a model of tolerance that
protects against the return of totalitarian narratives.

While Facebook provides the starting point for this discussion, the book's larger purpose and goal are to understand
Facebook society. In this context, it makes sense not only, occasionally, to look beyond Facebook but also to make excursions
into neighboring fields and themes and to revisit earlier eras in
the history of culture—philosophers and writers of past centuries and theorists of past decades. These excursions will offer
insights that also follow from Agamben's understanding of contemporaneousness through distance. One certainly could write
about Facebook society without mentioning Blaise Pascal, Gotthold Ephraim Lessing, Johann Wolfgang Goethe, Jean Paul,
Schopenhauer, Søren Kierkegaard, Walter Benjamin, Siegfried
Kracauer, Theodor Adorno, Paul Ricœur, Jean Baudrillard,
Roland Barthes, Zygmunt Bauman, Pierre Nora, Gianni Vattimo, Jean-Luc Nancy, and Judith Butler. But looking at the
phenomenon of Facebook together with these thinkers opens

up perspectives that point beyond the usual lines of argument. By adopting this perspective, *Facebook Society* opens up thought spaces that, even if they cannot be explored with all the patience they deserve, may inspire further consideration and future empirical studies.

FACEBOOK
SOCIETY

1

STRANGER FRIENDS

If ever I plead with the passing moment,
Linger awhile, oh how lovely you are!
Then shut me up in close confinement,
I'll gladly breathe the air no more

—Goethe, *Faust*, 1808

The character of an era is shown by its jokes. When I entered university, the shortest joke in the world was: "Two students pass by a pub." Today the opposite would be a candidate for first place. It is not that students no longer sit in front of beers, but in the era of mobile media and social networks, togetherness happens less often, or differently. And there is always the potential presence of the absent "friends," with their claims on those who are present. Spatial proximity is no longer a guarantee of intimacy when smartphones are lying on the table like pistols, ready at the next ringtone to transport the others into a communicative beyond. English has a neologism for this: "Phubbing"—a mashup of phone and snubbing. The term is already almost applicable in reverse. Is it not impolite to appear at a bar without a smartphone and in doing so to levy an implied accusation against the others when they turn to their internet business? However one interprets the situation, the shortest joke of the current era is probably "A person is looking around at his surroundings." And indeed, why should it be better, when you are waiting for the bus, to look at strangers instead of the updates on your social networks or news portals?

The friendship starts with Mark Zuckerberg, whose friend everyone is and whom no one can escape. Mark is the friend with the unlimited right to access all the pages of his two billion friends, including whatever they have long since deleted. He is the fomenter of friendships, who sends worried reminders if you haven't been heard from in three weeks: "Dear Roberto, You were not on Facebook recently. Here are a few people you might know on Facebook. Get connected with friends, family, classmates, and colleagues to view their recent news, pictures, and more." Mark is not unintrusive—"Your profile is 55% complete"—and not without his concerns—"thank you for stopping by again! We hope you will be back soon." Two months after your divorce, he manages, in his algorithmic networking logic, to suggest befriending the new partner of your ex-wife. You probably have to take him as he is, for it is not possible to "unfriend" him. He is as inescapable as family, more brother than friend, in other words, actual Big Brother.

"The Tyranny of Intimacy," the subtitle given to the German translation of a book by Richard Sennett that first appeared in 1974,[1] consists in leaving the other person no latitude to behave otherwise than he is expected to. Individual freedom results from social distance; its cultural-historical emergence came about in big cities, where people elude the social control of acquaintances and relatives. The internet, where originally you were only what you typed, initially seemed like an extension of the big city by virtual means, an "identity workshop" where you could playfully try on other identities. The most popular symptom, at the beginning of the century, was Second Life, where you could meet strange figures who might want to hear something from you but didn't know much about you.[2]

Subsequently, Facebook superseded Second Life and enriched the internet with the phenomenon of "ambient intimacy," namely, access to everyday details about individuals who are not actually known to those with whom their details are being shared. The two platforms embody very different communicative cultures, and precisely the contrasting types of self-perception they practice provide an important point of comparison. On Second Life, users saw themselves living via an avatar that conformed to the identity-theoretical concept of the 1990s, which was geared to experiment, flexibility, and play. On Facebook, users see themselves as living within a clearly structured and specified context. In place of the avatar in a fantasized costume, here are people with real names and photo IDs, framed by a design that is the same for everyone, without exception. This order, this "culture of real identity," was one of Facebook's advantages over Second Life and MySpace.[3]

Naturally, you can portray your life on Facebook as being more interesting, exciting, and mysterious than it is. People who are shy in communicative situations that don't involve technology can turn courageous when their selves are virtualized. But the frame doesn't leave much room for experiment. The friends, who are mostly also offline acquaintances, are witnesses to our selves who are hard to get around. The chronology ensures that the past remains present and threatens to reveal any contradictory descriptions of the self. The speed with which people typically bear witness to their lives hardly leaves time to manipulate postings, and both the entries by friends and the algorithms on a person's homepage expropriate authorship. Theater, if it still happens on social networks, plays out under conditions of strict observation, in an interactive communication process with responses (comments, likes, views) that serve as more or less subtle forms of social discipline.

There is something to be said for the assumption that Facebook is doing exactly what Zuckerberg has in mind when he criticizes the postmodern play of identities—even based just on business considerations: "Having two identities for your self is an example of a lack of integrity." Yes, the possibility to tag things as private—as shareable with "close friends," "friends except acquaintances," or "public"—still does make it possible, to a certain extent, to represent yourself differently to different people. But this is the extent of the latitude that users have wrung from Zuckerberg, who would prefer to stick to the basic setting "public" for all postings. The person who ignores you on a big-city street is Facebook's nightmare; the village on a global scale, its paradise. Everyone, at least potentially, knows one another too well to pass by anonymously. That Second Life, nowadays, doesn't even ring a bell for many people shows how much the internet has changed. Apart from oases of anonymity with questionable contents, like World of Warcraft or 4chan, ID is required. There is no second life in the first one. The "avatar" on Facebook has the same name, age, appearance, and profession as the person behind it. When teenagers list their age as "69" (as a symbol for double oral sex) and their place of residence as Afghanistan or Zimbabwe,[4] they are not pretending to a false identity but making insider jokes that even outsiders can see through.

Life on Facebook is by no means as wild as the pioneers of the 1990s imagined the internet of the future would be. From their perspective, the success of ordinary life in Facebook's restrictive format is capitulation a billion times over—the betrayal of a life that should be special by one that is especially controlled.[5] But why, then, is Facebook able to captivate millions of new users every day? Because it provides the technical answer to a cultural problem? Because it vanquishes postmodern uncertainty—which Second Life playfully stages—and replaces it with reliable transparency? Because it is the perfect storeroom where the restless, homeless individuals of our era, who no longer find a firm foundation in the things they encounter, can dispose of what happens to them? Facebook, unlike Second Life, is not a place where a person can have adventures; it is a place where a person can narrate his life *as* an adventure. Facebook is the perfect therapy for *horror vacui* in times of incessant experiences.

FRIENDSHIP IN THE TWENTY-FIRST CENTURY

The nature of the human brain is such that an individual cannot handle more than 150 social relationships. This was the result of a study carried out in the 1980s by the British anthropologist Robin Dunbar. Ever since Facebook has been in existence, there has been controversy over who is wrong, Dunbar or the relationship concept on Facebook. Zuckerberg is of the opinion that Facebook broadens the "social capacity" for "empathetic relationships" far beyond Dunbar's limit.[6] But anyone who lavishes his critical energy on a critique of the phenomenon of friend collecting on Facebook or seriously thinks a thirteen-year-old girl is hopelessly swamped by her 450 Facebook friendships runs the risk of being laughed at, and not only by teenagers.

Naturally, people in many countries, no matter what language they speak, know that on the world's largest social network, the word "friends" is a "false friend," as foreign-language teachers call the foreign words that only sound like what they appear to mean—like German *brav*, which does not at all mean "brave"; *Gift*, which means poison, not "gift"; or other similarly misleading examples. The difference was more clear when it was still acceptable to use artificial terms like "Friendster." On Facebook, the word is "friend," again. Usually it does not mean more than "acquaintance," and often much less. Sociology, meanwhile, has come up with paradoxical expressions like "intimate strangers," "anonymous friendships," or "weak ties."

His gaze has grown so tired from passing through
A thousand messages, it now holds nothing more.
There seem to be a thousand friends on view
And after a thousand friends no world at all.

This is how Rilke's poem "The Panther" sounds in the twenty-first century, as the template for a critique of "friendship society."[7] In the flood of updates, the subject wearies; its friends are keeping it from the world. Could it be that the friends on Facebook are not so much foreclosing the world as promising to save us from it?

The flattening out of relationships is one of Facebook's foundational ideas. It was invented, after all, as a protest against the exclusivity of Harvard University's clubs, which had refused entry to Zuckerberg, the offspring of ordinary parents. The less "friendship" on Facebook focuses on content, the more it is determined by numbers, the more inclusive it can be. This openness also distinguishes Facebook from MySpace, which targets Generation Y, mainly in the realm of pop music. Facebook's declared mission, to connect the world and thus secure world peace, is ensured above all—this is its deeper meaning—precisely by the superficial nature of its relationships. The very lack of substance and more ambitious claims that characterize friendships on Facebook help create links that transcend shared values, interests, or ideologies. The identity of differences on a thirteen-year-old girl's friends list is pluralism in action. The strength of "weak" relationships consists in the fact that their subjects are linked by nothing other than these relationships—relationships for their own sake, a disinterested delight[8] in communicating.

This is not only true online. Facebook's imperative to connect has long since changed the way we relate to one another outside of digital media. Who, ten or fifteen years ago, would have given his telephone number to the casually encountered acquaintance of an acquaintance as freely as people, nowadays, "friend" someone after exchanging a couple of sentences on Facebook? But once you are connected, you are also exposed to the other person, to her photographs, updates, and opinions. The receptivity to new relationships is supplemented by a continuing openness to old ones. When Facebook urges someone to connect with former classmates, it is treating friendship like a family relationship: Individuals are linked to one another not because they share the same opinions but because of a common origin—in this case, a shared past rather than shared blood. The time that was once spent together bridges temporal gaps, geographic boundaries, and intellectual, cultural, or political differences. It generates a current bond that would not exist without the technical and social *dispositif* of Facebook.[9]

This quick glance at history suggests that the friendship model on Facebook acts as a corrective to a nineteenth-century development that led to the decline of previous models of collective sociality. Friedrich Schleiermacher, in his 1799 "Essay on a Theory of Social Behavior," describes the problem as a result of the specialization that the social life of modern humans necessarily entails. "A profession restricts the activity of the mind to a narrow compass. However noble and admirable it may be, it always holds the impact on the world and the views of the world to a single standpoint; thus, the highest, most complicated profession engenders the same one-sidedness and narrowness as the simplest and most humble."[10] The solution to this problem, for Schleiermacher, lay in the social circle, as a form that could temporarily suspend the "constraints of domestic and civil relationships" by intersecting a person's sphere "with the spheres of others, as variously as possible . . . , so that all the phenomena of humanity gradually become known to him and he can become friendly and neighborly, as it were, with even the most foreign hearts and relationships."

What Schleiermacher hoped for at the end of the eighteenth century was no longer to be expected in the first half of the twentieth. Society had become so differentiated that its representatives could hardly be thematically unified any more. The only thing that appeared as a "universal human element, . . . what they all have in common," wrote Robert Musil, is "stupidity, money, or, at the most, some leftover memory of religion."[11] Seventy years later, social networks offer a fourth possibility of "generally human" characteristics that transcend specialized professions and particular interests. The urge to communicate, often quickly dismissed as a mixture of narcissism and voyeurism, creates a sociality of differences, which since the Enlightenment has been an object of modernity's wishful thinking.

This use of new technology to solve an old problem is not without its contradictions. For Facebook, too, has its thematically specific groups, and the statistical logic of its algorithms leads it to create a "filter bubble" that orients communicative relationships in ways that tend to confirm and strengthen existing interests.[12] But this does not alter the in principle possible and in reality often quite accidental mixture of Facebook friends. Even if Facebook cannot prevent the continued existence of like-minded communities, its logic of maximal friending points beyond the traditional model of the creation of exclusive groups. Nowadays, you are exposed to the contrary opinions of a former schoolmate because Facebook suggested a relationship that would otherwise not exist—regardless of whether this person's opinions have meanwhile evolved in directions fundamentally different from your own. The ever-present internal filter of our communicative decision making has become porous; the technical *dispositif* of Facebook (the "filter bubble" of its algorithms) is corrected by its social *dispositif* (the imperative of having a large number of "friends"). Representatives of all the periods in a person's life (schoolmates, college friends, ex-partners), of all the self's social spaces (colleagues at work, sports team members, relatives) come together in an "individual melting pot" that, as empirical studies show, does quite well without political agreement among those who are connected in this way.[13]

The ease of "friending" in the relationship model of social networks is often viewed as negative on account of the lack of commitment that comes with its immediacy. In Germany, before the success of social networks in the 2.0 era of the World Wide Web, critics of virtual communities saw their often-emphasized benefits—the equal chance for participation by those who would otherwise be disadvantaged—as a loss. For example, say the detractors, if the phenomenon of disability is covered up during communication, there is no need to develop an appropriate response. The conclusion that "in the virtual world, social difficulties are eliminated without there being any necessity to develop [corresponding] social capabilities" is followed by the culturally pessimistic prognosis of a worrisome spiral: "Capabilities that are not used wither away. As a result, real life becomes even more difficult. This, in turn, increases the attractiveness of virtual reality."[14]

This perspective persists today when the conversation turns to relationships that are not associated with any cost, friction, or risk. The distance created by the screen, critics aver, allows a control of the communicative situation that would be unthinkable in real physical proximity, including the possibility of turning the other person off at any moment. The challenge that normal social relationships bring with them—so the argument goes—is passed over; there is no need to engage others and allow them to call you into question. An interpretation that goes even further argues that the wish never to be alone yet always to be in charge is already preparing the way for life with a robot, which promises precisely this low-maintenance relationship.[15]

Talk of the decline of friendship, even when it comes from the trap of "false friends," is in fact quite justified. The modern relationship model is operatively based on noncommitment, and communication broadcast to a disparate and widely dispersed group of friends, with no specific address, is less personal than the advertisements the recipients consume along with it. And even when a personal salutation is included, atypically, it often goes out through a metaphorical megaphone. At the same time, we can give the situation a positive spin if we wish. For one thing, public endearments like "Miss you" do not necessarily have an unpleasant ring in the ears of their recipients. For another, the lack of commitment can be seen not as a flight from responsibility but instead as a posture of openness to options, as a triumph of space over time, the shift from a syntagmatic to a paradigmatic model of life. When changes in life circumstances bring new friendships with them, the time and attention that the traditional friendship model demands can require radical breaks. Facebook, on the other hand, makes it possible to remain in low-level contact with friends from earlier periods in a person's life. The chronological logic of the syntagma gives way to the paradigmatic abundance of candidates for the category "friend." Chronological change does not require spatial change: no moving on without keeping on. The social network becomes an archive of social relationships that can be more or less reactivated and intensified at any time and that often come to mind again only when the system announces the other's birthday—which is why a frequent "birthday present," brutal but consequential, is the message ending a Facebook friendship.

This open stance corresponds to the "liquid" identity concept of the postmodern subject, which liberates itself from petrified relationships as it also maintains its distance from commitments. For the preferred "easy-to-enter and easy-to-exit" relationships, the social networks of the Web 2.0 provide the appropriate technology, which in turn—as the reification of social facts—cements this relationship model.[16] With this, society has moved as far as can be imagined from the Romantic model of friendship, whose unconditional variant was offered by Friedrich Schiller in his Hollywood-style ballad during the heyday of German classicism (long before Hollywood's 2003 animation *Sinbad: Legend of the Seven Seas*): "The Hostage."[17] The friend of an unsuccessful tyrannicide offers himself as a hostage so the guilty man can marry the friend's sister before he is executed. If he doesn't return within three days, the friend must die, but the conviction will be lifted. The return is threatened by flood, highwaymen, and exhaustion. But the guilty man gives his all and returns to the scaffold on time, whereupon the tyrant, deeply moved by such friendship, asks to be accepted as the third member of the band.

Naturally, dying for your friend was not part of the standard model of friendship in either ancient Greece or German classicism. But that true friendship proves itself in adversity was a platitude even before Aristotle, along with the notion that friendship is exclusive and limited, as Schiller noted in his poem "Friendship": "Joy! Joy! I have found you, embraced you among millions, and among millions, you are mine." This minimalism is the result of a realistic weighing of resources, according to which the richness of a single relationship demands a poverty of relationships overall. The laws of attention economy apply to friendship, too.

It is not without paradox that the information society, in which attention is such a scarce and valuable good, aims to maximize the number of friends. The result is a model of friendship whose investment strategy defies all Romanticism. Whether it's WhatsApp in the subway or Facebook in the elevator, something is always possible. Experts can easily post twenty likes between one stop and the next. That an original comment by an introverted, occasional poster has fewer chances than the visual banalities of a popular "liker" will astonish only the nostalgic. The business of liking is opportunism under time pressure; we disburse likes not for quality but for likes—assuming it doesn't take too long. Critics hope that by overcoming quantification we will see the return of the listener, if not of the friend.[18] But as long as there are still numbers on Facebook, there will also be friendships for sale and enterprises with brazen names like Socialyup.com to provide the desired commodity. People who turn to inexpensive variants will only wind up with the friendship of bots, which dispense likes with telltale heartlessness: five hundred for $30, three thousand for $130. Attention and comments from living persons are more costly. Could a "false friend" of this kind ultimately turn out to be the most loyal companion? Let us imagine the following:

A student in Vladivostok—let's call her Natalia—takes a job, via an advertising agency, as a Facebook friend of Jenny Doe. Her job: not just to post likes on Jenny's page but to add clever commentaries (beyond "great," "cool," or "love it") on almost everything Jenny writes (update, life event, comment), reads (weblogs, news sites), and sees (YouTube, Vimeo, Instagram). Natalia is expensive; like Jenny, she has a PhD in literature. Natalia follows Jenny to every site on the internet. She does it gladly, for she likes Jenny's taste. Thus, late in the evening, when her friends are already sitting over vodka, Natalia can often be found in front of her screen, still on Jenny's trail, in Jenny's world. Deep into the night, she searches for formulations that could please Jenny. She wants to earn her keep; she wants to make Jenny proud to have a friend like her. When Jenny posts a comment in response to Natalia's comment, the smile still lingers on Natalia's lips as she sits with the others over vodka. Is there anyone who can't predict how this the story ends? When Jenny stops paying, after a couple of weeks, Natalia asks to remain friends even without payment, happy to have found, among millions, the one person with whom she feels an elective affinity. Jenny is touched, but more confused than touched, and immediately breaks off all contact with Natalia.

What potential for real friendship! What material for the long-overdue Facebook novel, *The Friend You Bought Yourself: Pleasure Her to Possess Her*. This stranger-cum-Facebook-friend would preserve the model of romantic friendship for the twenty-first century: on Facebook against Facebook.

SELF-REPRESENTATION AS CONSCIOUS CONTROL

When YouTube, in 2006, replaced the self-description on its welcome page, "Your Digital Repository," with the command to "Broadcast Yourself," it joined the self-realization slogan of the 1980s, "Experience your life," with the self-exploration maxim of the 1990s, "Narrate yourself."[19] Facebook does something like this when it invites us to share our own life with others, day after day, by posting not just biographical transitions but all events, significant or insignificant. Social networks are both biotope and stress test for "Generation Me," for whom the compulsion to have something to tell makes hedonism more or less a duty. The person who has no attractive events to report feels socially disqualified. What previously went unnoticed now creates a noticeable gap. Surely, it was to be expected that one consequence of the self-representation imperative would be manipulated reporting—and that the results would include depression (when your life pales in comparison with the brilliance of others) and self-deception (when you later remember false claims as having really happened).[20]

Since language is never neutral, it comes as no surprise that the managers of social networks operate with terms like "transparency," which has a positive connotation, and in this way suggest that the obsessive publication of one's life should be understood as a social gain. Zuckerberg's famous mantra that more transparency means a better world is also, he claims, the reason why the desks in Facebook's main office are located in a huge, open hall; even the conference room is only separated off by a wall of glass. The not always ironically used rhyme on the dogma of transparency is "sharing is caring." The antithesis necessarily follows, with its critique of reticence as asocial behavior: "privacy is theft."[21]

The conceptual upgrade of transparency is opposed by a corresponding conceptual downgrade, for example when an exaggerated drive to communicate is discredited as exhibitionistic and narcissistic. Sweeping generalizations of this kind ignore the historical roots of the concept of transparency in art and social utopias. In the context of the "glass culture" of the early twentieth century, both communist ideology and avant-garde art introduced "glass man" as an avant-garde, alternative identity concept that could be wielded as a weapon against bourgeois culture. In the latter half of the twentieth century, the culture industry may have appropriated the concept of public privacy, but it resurfaced, at the turn of the twenty-first, in the form of individual experiments on the internet that were directed against mainstream culture—before this new round of "subversive exhibitionism" could also be taken over by the social networks and turned into a tame, mass phenomenon they could exploit.[22]

Narcissism also has a subversive prehistory. In the late 1940s, psychoanalytic theory already understood it not simply as the revolt of the id against the ego but also as the revolt of the ego against an alienated world. It was Wilhelm Reich versus Sigmund and Anna Freud; the orgasm versus neuroses. The opening toward the id was made in the name of a richer and more mature ego that could successfully escape coercion by the old order. By the 1970s, at the latest, narcissism had become a positive component of the self, which ultimately provided the basis for diverse emancipation movements: youth, women, homosexuals, the

New Left. To dismiss the demise of modesty, as symbolized by excessive self-representation on social networks, with Andy Warhol's sentence about fifteen minutes of fame would thus be to miss the cultural-historical tragedy of the historical event that transpired here. The liberating blow of self-expression, as the alternative to unsuccessful political activism, was finally integrated into the process of capitalist exploitation as a nonconforming, lifestyle-specific form of consumption. The Enlightenment imperative of coming of age wore itself out in the self-expression of consumers. Within the framework of the social networks, self-representation now even stands under a triple sign of consumer culture: as a badge of the capacity to consume, as the basis for personalized marketing of consumer goods, and as psychological conditioning for the acceptance of this very social constellation.

If narcissism is the source of exhibitionism, voyeurism is its possible consequence. What is exhibited can be observed. Making private things public is naturally inseparably intertwined with the debates about surveillance, which, in turn, are supported by two central arguments: relativism and reciprocity. Privacy, the representatives of the postprivacy perspective assert, is not a universal and timeless value but a social construction of Western culture since the eighteenth century.[23] The argument that it hasn't been around that long is somewhat absurd if one considers the other achievements of Western civilization, whose relevance is by no means in doubt on account of their relatively brief duration: democracy, freedom of opinion, women's suffrage, and many very recent rights for minorities. But the loss of privacy was already something people could support before September 11, 2001, if it promised an increase in justice. Already in the 1980s, the ability, via the mass media, to look into the private lives of public figures, including their moral failings, such as tax fraud and adultery, was presented as an act of enlightenment that exposed the "'ordinariness' of everyone."

The accompanying warnings were not about the loss of privacy but about the asymmetry of the loss if those who were doing the observing could not, themselves, be observed.[24]

This very danger of unequal observation is growing today, after society, in the form of social media, has created a space for itself in which *everybody* can be ordinary in public. The danger comes from unequal opportunities for access to knowledge as it is available and subject to analytic focus online. The new digital divide lends new pertinence to the concepts of the Panopticon and Big Brother, which had seemed inapplicable thanks to the reciprocity of transparency and surveillance first in the mass media and then on the internet. If, today, Facebook serves as the staging ground for new forms of surveillance—this is where the software for facial recognition or personal identification based on posture is being developed and tested, with the help of playful tagging by friends—then it may be true that, in the first instance, everyone who moves within this network is affected. But if the back end of the interface, the "other" side of the website, which is blocked for "normal" users, makes distinctions in users' capacity to acquire data based on rights of access, data-collection resources, or analytic competencies, then it no longer makes sense to talk about an equal observation of everyone by everyone else.[25]

Critical and Marxist-inspired theory, for this reason, quite justifiably regard the social web as the most important source of data "for the creation of knowledge to be used for governance and control." This includes knowledge for *commercial* domination, as personal interests are analyzed with the aim of effective advertising. The debate over the political and economic capitalization of data circulation is carried on under rubrics like dataveillance, biopolitics, and governmentality, as well as free labor, self-branding, and gamification.[26] A central theme, at a higher level of abstraction, is the shift from discursive to algorithmic control of social processes: to "cybernetic" or "algorithmic governmentality." Just as, in the theory of governmentality developed by Michel Foucault, social control is imposed not by making a topic taboo but by locating it within a specific "discursive regime," the strategy of governance via certain narratives is giving way, with the digital media, to a governance strategy in the numerical mode. The reason for this is to be found not only in the nature of the computer, which is a logic of databanks and computation, but also in the nature of the internet, whose pluralism of opinions and tangle of voices undermine the status of governmental institutions—agencies, schools, mass media—as central platforms for official statements. "Pastoral power," the name Foucault gives to techniques of discursive control (including structures outside the church), loses its listeners in the latter's uncontrollable, networked communication; the increased production of knowledge available for control compensates for the decrease in control over knowledge.

At the same time, under the cybernetic regulatory regime, open, ambiguous, and undecidable communication is being replaced by "mathematized communication," which operates according to a logic of decision making that is also "discretized" and "binarized." Administration theory might characterize the replacement of politics by mathematics as "statistical democracy" and might even welcome it, especially in light of the difficulty of keeping order in globalized mass societies. "Statistical democracy" could denote a society in which normativity (think of adultery or homosexuality) is determined by the *statistical* existence of a given issue and not by its ideological or moral status. But from the perspective of political theory (and not only political theory), the replacement of narration by numerical representation, of argument and conversation by numbers, of a culture of discussion by the immutable decisionary logic of algorithms is highly problematic—quite apart from the accompanying move away from the offensive indoctrination ("information") of the population and toward specific interventions in discrete contexts and actions. While cybernetic governmentality is not the theme of this book, the aspects of Facebook that are addressed here must ultimately be understood within this broader political context. The loss of reflective perception of the world and the self, at the front end of the interface, is only a different form of what is taking place at its back end, in the form of the cybernetic formalization of reflective processes as algorithmic analyses and rules. In each case, what is occurring is a reduction of communication, in which the elements of the discursive, the narrative, and the linguistic are lost. The frequently cited shift from narration to the model of the database as the predominant paradigm for meaning making is taking place in many different forms. What will be discussed below, in the context of reflection on social networks, is ultimately also a contribution, drawing on current practice, to the theory of a future cybernetic governmentality.[27]

The familiar objection that Facebook is nothing but a waste of time that clogs the public realm with individual banalities is basically a rehearsal of the central claim of the theory of the culture industry: stultification through mass culture. According to this theory, distraction and amusement are tools for securing political control. Adorno is convinced that to be amused is also to be in agreement. "The liberation that amusement promises is that of thinking as negation." Critical theory held the culture industry, which it claimed had been able to implant a false need in the masses, responsible for the fact that most people have no wish to negate the social status quo. From this perspective, the masses, with their poverty of reflection and their addiction to pleasure, are blind. They can't tell the difference between appearance and reality; they are "sick people, whose sickness consists in the fact that they don't realize they are sick." Since reception theory and cultural studies have since described the relation between the public and (pop-)cultural artifacts as substantially more complex, as well as more resistant to manipulation, it is no longer possible to speak this way about the audience for products of the culture industry. Is it possible to talk this way about Facebook?[28]

The answer is neither a simple yes nor a simple no. It must begin by clarifying a different question: Is Facebook merely the implementer of a cultural trend, or its actual driver? Zuckerberg's statement that he is only providing the technological means to implement new social norms is initially only a defensive claim. It is part of the rhetoric of tech companies to describe the internet as the motor "driving one of the most exciting social, cultural, and political transformations in history," which will raise "fundamental questions about identity, relationships, and even our own security." At the same time, however, the firms prefer to identify the users of their technology, not the company or its products, as the driving force behind this process.[29] In concrete terms, the thesis that Facebook only provides the technical realization of new social norms is not defensible, considering that Facebook has repeatedly imposed or attempted to impose innovations without any need on the part of its users, and partially against their will—think of Newsfeed, Beacon, or Timeline. The company was criticized repeatedly by its users for this and has had to apologize and reverse some decisions. All of which changes nothing in its tactic of taking two steps forward and one step back. Zuckerberg is much too convinced of his own role as a visionary to be satisfied with meeting needs; he is someone who anticipates and creates them.

And yet: At the same time, Facebook did emerge from a historic need, a need for the very cultural values that the new technologies (help) create. Would Facebook be as successful as it is if it were merely attempting to persuade people of something they didn't need? The usual reference to the "lock-in effect," with its claim that Facebook is too powerful for anyone to escape it, is as unsatisfying as Adorno's argument that the need for the products of the culture industry is already an expression of the latter's successful manipulation. Bogeymen like these are self-serving simplifications that serve only to invite the consoling idea that a better world would be possible if only we could successfully remove the source of the problem. Without the culture industry, so this argument runs, human emancipation and social democratization would be better off. Similar formulas are being repeated nowadays in regard to Facebook, whose model of society is said to betray the emancipatory and democratic potential of social networks in favor of the profit interests of the company's owners. The upshot of this critique is, then, frequently a demand for alternative social media that operate without commercial objectives and in line with strict data protections.[30]

To avoid giving a wrong impression: It is true that a *for-profit* network like Facebook is setting the parameters for an "advertising and economic surveillance machine" and that the parameters it sets for understanding the self and the world are aimed at the commercialization of social communication rather than the emancipation of human beings. Facebook's business model contains the "original sin" of every social network. From it, there necessarily follow decisions that are not centered on the interests of its users. Facebookers are products of Facebook in a dual sense: as organized attention that is sold to advertisers and as subjects formed by Facebook's technical and social *dispositif.* In the terms of the Marxist design of critical theory, the economic being of the social network determines the consciousness of its users. This constellation is impossible to ignore. It follows that the accusation of distraction and manipulation is also understandable. Critical theory had turned Kant's theory of knowledge into a social theory by deriving the structure of experience and consciousness not only from the concepts that were available but also from the real social constellation. In the era of cybernetics, the categories of knowledge must also be brought up to date with reference to the technological *dispositifs* of social life. This occurs in those software studies that focus on the underlying programming. A media theory that is interested in cultural theory must inquire, first of all, into the form of self-knowledge that the photographic and algorithmic method of self-representation makes possible on social networks.

At the same time, it is also important for the inquiry to go beyond the political-economic perspective, with its sweeping account of deception and manipulation, and to dig down to the anthropological core of the problem. It is, even then, still possible to understand Facebook as the expediting symptom of the publication of private life and the flattening out of communication and to understand it *only as* a symptom, with the motor localized in the *conditio humana*. At stake is the psychological situation of the modern subject, the question of the extent to which a person communicating on social networks gains a sense of the spiritual home she has lost in the process of civilization. Just as those surveillance studies that have begun to discuss surveillance strategies as a positive social practice that goes beyond the mantra of institutional misuse are doing,[31] it makes sense to discuss Facebook not (only) as a huge data sweep but (also) as a "savior in adversity"—an adversity that becomes clear as soon as we move from describing the social networks to reviewing their historical and philosophical context.

COMMUNICATION AS FLIGHT FROM THE NOW

The weapon of every tourist is the camera. This is how tourists enter every church, temple, and museum, every viewing platform in a national park. Shielded by their camera, they scurry on, unseeing, as soon as the necessary minimum has been accomplished. In the service of future memories, they ruthlessly toss away the present and sacrifice the dignity of seeing to the archive. In the flashbulb storm of the visiting group, the object of visual desire loses the last little bit of aura that remained to it. Visitors who have managed to shoot a photo behind the back of the guard hasten onward with a gleefully impish expression, as if the same photo did not exist on the internet, freely available and a hundred times better quality. "Standing face to face with one of the great wonders of the world (let us say the *patio de los leones* of the Alhambra), the overwhelming majority of people have no wish to experience it, preferring instead that the camera should."[32] What the Italian philosopher Giorgio Agamben described in 1978 is true, by now, even of rock concerts, which many fans, in the expectation of a good photo for their friends on Facebook or Instagram, now follow only on the little screen of their raised smartphone. Can one not experience more without a medium? Can we still experience anything at all?

The proofs that are borne back from There and Then bear witness to the missed Here and Now. So you were there! Then you could have seen this: the smile of the Mona Lisa that has looked into thousands of eyes! The perhaps really final Rolling Stones concert. But for looking there was no time and, above all, not enough courage. Yes, courage, for the eagerness of the photographer conceals more than ignorance or the intent to provide evidence of the place we have been to. There is also the fear of the object. Of standing in front of it breathless, unmoving, contemplative (as people used to say). Of being regarded and reminded by it. "There is no place that does not see you," wrote Rainer Maria Rilke, beseechingly, in his poem about a famous Apollo statue, leading up to the celebrated last line: "You must change your life." What if you don't know what to say to that? What if you feel the pressure, or the emptiness? "See Venice and die," we are famously commanded. What an absurd saying, in our "experience society." Whether it is Apollo, Mona Lisa, or St. Mark's Square, the moment's aura vanishes when confronted by passionless satisfaction and the dinner plans that follow the picture taking.

When we are unable to grasp something, we hold on to it. What we can take home as a photograph is exorcised. That there are better photos on the internet is no argument when what matters is not having the most beautiful photo but that, as Kafka suggested, "We photograph things in order to drive them out of our minds."[33] The defense would argue for the photograph as an occasion for thought, with whose help the photographer will later do what he has no time or desire to do in the experiential moment, namely, construct a personal relationship to this object, this situation. Since, however, it often does not come to this even after some time has elapsed, Kafka's remark holds true—with the additional note that time and quantity have an impact. For it is one thing if we consider the case of the person who brings only two or three images home from vacation, buried in a roll of thirty-six-exposure film, to which on the next vacation three more photographs will be added. What a surprise, years later, when the film is developed! What an experience of recollecting—how he returned surreptitiously, each day, to this motif that had immediately struck him; how, shortly before departing, he finally took the camera along. What insight, now, as he studies the photo, into why it was so important to him then.

Without the increase in value that results from this reduction, the photo is a betrayal of the present to the future. The person who only perceives things through the camera is always acting as the advocate of a future viewing. At least, this was still true when there were still photo albums or slide evenings for examining the booty that had been seized from the past. Even this hardly amounted to a salvation of the missed moment, but at least the accompanying story made the experience somehow whole. Today, when there is a camera in every telephone and the slide show brooks no delay but instantly—minus the narrative effort—appears on the social network, the time for future recollection has gone missing. The archive of images fills up too quickly for us to have the energy it takes to return to them. The more photos we take, the less we see.[34]

As for being filled up, technology is once again attempting to solve the problem that technology brings with it. Apps compress the past by creating a film made up of the best second of every day (1secondeveryday.com) or remind us to be reminded, by pointing us, every day, to the photos and updates (on Facebook, Twitter, Instagram, etc.) we took exactly a year ago (timehop.com). Other apps provide training in storytelling in the form of "visual stories" (photos with textual commentary), which can be shared on various platforms (storehouse.co, steller .co). Memory aids like these present themselves as the solution to an organizational problem. They ignore the fundamental motif that Kafka and Agamben identify in the urge to document the present. Agamben's formulation should be taken seriously: We prefer to let the camera experience what we are going through. This is not about postponing perception to a time in the future but about delegating it to other authorities.

When it came to putting perceptions into storage, the camera was never that effective. Left to machines, "delegated enjoyment" runs empty.[35] What is needed are addressees who have an equal capacity to perceive, which is why there was ultimately the slide show or the hope of being able to show the images to someone, sometime. One can even take refuge in children, which results in surplus pedagogical value: "Look, a rainbow"; "That's the famous Mona Lisa"; "The Grand Canyon is more than five thousand feet deep!" Social networks like Facebook, Instagram, and Twitter make it possible to delegate enjoyment to partners of equal value, without incurring any time lag at all. The trick of digital media lies partly in the cost-free nature of the photos, which encourages more intensive use of the camera as protection from the demands of the present, and partly in the pact with the social, which lets the delegating of experience appear as communication rather than repression. Facebook is the logical consequence of the situation diagnosed here: It destroys the present by holding onto it. The next step in this coping strategy is Snapchat, which goes so far as to abandon the archive.[36]

The delegation of enjoyment aims at a double salvation: from the challenge of the object as Rilke conjures it up and from the lack of passion that Agamben supposes exists "face to face with one of the great wonders of the world." Delegated enjoyment does not signify the negation of pleasure but the fear of it. This is not the fear, or wiliness, of Odysseus, who according to Homer had himself tied to the mast and stopped the ears of his crew in order to get past the Sirens without damage—a form of disciplined enjoyment whose spirit would later inform the invention of fat-free whipped cream and nonalcoholic beer—but fear of the sense of inadequacy, the void you might feel internally, to which the challenge "You must change your life" could be addressed; or the fear of the silence of the Sirens that Odysseus experiences in Kafka. We escape from this feeling through the busyness of communicating on the social network, which demonstrates our capacity to act in the moment of helplessness.

Agamben refers to Benjamin's reflections on the loss of longer-term experience (*Erfahrung*) and its replacement by merely incidental lived experiences (*Erlebnisse*) in the essay "Experience and Poverty,"[37] which will continue to occupy us in the following. For the moment, we will confine ourselves to Benjamin's information about how the problem was being solved at that time: by the souvenir that was brought home from the scene of the event, as complement to "the defunct experience which thinks of itself, euphemistically, as lived experience."[38] The souvenir provides an "honorable" form of liberation from experience, through which people may "make such pure and decided use of their poverty—their outer poverty, and ultimately also their inner poverty—that it will lead to something respectable."[39] Photography becomes an expanded form of "respectability," which, as something a person herself has created, can even be personal. On a social network, ultimately, collective feedback elevates this act of personal communication. It may rarely extend beyond shares, likes, and one-syllable commentaries ("cool," super," "wow," "envy"), for the social network is subject to laws of the attention economy. But the communication fulfills its function all the same, by making the current moment part of the time of the social network in a threefold process: first, by taking the photograph at the original location; second, by feeding it to the network; and third, by responding to the feedback that starts as soon as the image is uploaded and subsequently by responding to the status reports of the others.[40]

Incessant reporting on oneself turns out to be the best defense against oneself, as we send the moment of lived experience "home" to the network. The camera is not an apparatus for seizing spoils; it is the shield we hold up to avoid taking even a moment away from the busyness of communicating. Documentation of the present takes place in the interest of getting through it, not of remembering it at some future time. Here, the concept of "present shock" must be taken as seriously as Douglas Rushkoff did, in the book of the same name, when he described present shock as the "state of constant distraction in which we can no longer distinguish what is unimportant from what is important" and went on to observe that "Instead of finding a stable foothold in the here and now, we end up reacting to the ever-present assault of simultaneous impulses and commands."[41] It is this constant barrage that sucks us dry and that we can no longer live without. The present is not only no longer enough; it is also always too much, thanks to its nonrelation to the rest of time. The "tyranny of the moment" lies in the sequence of moments that, no matter how intensive they may be, mean nothing. The lack of passion is something we experience not only when we are confronted with works of art or the forces of nature but in our everyday lives, as well, which have assumed an urgency that is continuous and at the same time completely banal. The restless apathy, the sense of being lost in a formless sea of events is salvaged by sharing this Here and Now with friends on the social network. We are catapulted out of the lived, experienced moment into the communicative parallel world of the social network: technologies of sharing as technologies of self-avoidance.[42]

The medial center of this flight from experience is the photograph, as that element of the social network in which the delegation of the moment to external authorities is materialized. The photo, consequently, holds a melancholy different from the one we find in Roland Barthes. For Barthes, photography was a melancholic medium because every photo is potentially a document of death: It presents a Here and Now as a more or less distant There and Then. On the social network, the photograph loses this impact because of the shortened span of time between the moment when it is taken and its viewing. The photograph becomes less a document of death than a not-becoming-alive, since it represents the moment that is captured less as something that once existed and is now lost than as something that has been refused. The "Kodak moments" that Kodak's campaign promoted in the 1970s were moments that belong to the camera not because (as Kodak saw it) the camera is holding onto them but because (in the view expressed here) they are being abandoned to it: delegated to, "experienced" by the camera.[43]

This betrayal of the present no longer happens in the service of something that will become present once more at a future time. What happens, instead, is that at the moment of communication the individual finds herself simultaneously inside and outside of all three forms of time. If, in earlier periods, people wrote about the past in a Now that anticipated a future reading, now our social networks collapse all three temporalities into one. What we experience is simultaneously captured and perceived by others. All three temporalities coincide on the social networks. The permanence of lived experience (as an imperative of the experience society) assumes (under the imperative of self-representation) the permanency of a report. Life is lived in the form of sharing. From this perspective, the obsessive self-representation on social networks is an expression not of vanity but of suffering—and of solidarity, which we experience via the brief but certain attention of the others. The social network proves to be a community of need, adversity, or affliction, a "machine" for dealing with the present, in which each act of communication responds to the lived experiences of the others and—a "group cuddling" by likes—"takes care of" their experiences as a kind of labor of love. Before taking a closer look at the labor of love, however, let us take a short detour to explore the origins of the need.[44]

Young people today seem to have no need of being alone and cannot imagine why anyone should have such a need. Instead, they use the new technologies as a permanent defense against this danger. This is a broadly accepted finding that occasionally ends with a warning: "If we are always on, we may deny ourselves the joys of solitude." Even if aloneness plus insight (solitude) is clearly distinguished from depressing aloneness (loneliness), its attractiveness is hard to communicate to the "digital natives." On the contrary, the connectedness of social networks and mobile media guarantees a permanent sense of security in a virtual society, as internet theorists confirm: "We initially love them [the social networks] for their distraction from the torture of now-time. Networking sites are social drugs for those in need of the Human that is located elsewhere in time and space." It may remain unclear in what the torture of now-time consists, but the reference is relatively easy to supply, for what we have encountered here is an old theme of modernism.[45]

In 1948, the Swiss cultural philosopher Max Picard, in his book *The World of Silence*, complained about the loss of silence. The complaint was directed at radio as a mechanism for pointless communication, one no longer concerned with content and instruction but instead purveying "pure word-sounds." The term of art for this finding—"Radioitis"—comes from the 1920s, when essays like "The Proper Diet for the Listener" warned against the excessive consumption of radio. Picard's primary theme was not media critique, however; it was the cultural changes that came along with the mass society of the twentieth century. His real topic, as expressed in an earlier book, was *The Flight from God* (1934) and the resulting loneliness of human beings, for which the radio, as the most current medium of distraction, provided an effective mass antidote. It should come as no surprise that today the very same media critique is being disseminated by no less than the pope himself, with reference to the internet and social networks.[46]

Picard repeats a figure of thought that derives from the French philosopher Blaise Pascal, for whom it was already clear in the middle of the seventeenth century that "all the unhappiness of men arises from one single fact, that they cannot stay quietly in their own chamber." Left to himself in solitude, a person would presumably reflect on the hardships of life and on his mortality. Even the king, "if he be without what is called diversion . . . is unhappy." Thus he would not want to have the hare he is hunting given to him as a gift, for the main thing is to kill time. The alternative to both the hunt and distraction is refuge in God, which Pascal promotes with his famous "wager."[47]

If this sounds antiquated, in 1985 the French philosopher Giles Deleuze offered a version with a more critical accent on power: "It's not a problem of getting people to express themselves but of providing little gaps of solitude and silence in which they might eventually find something to say. Repressive forces don't stop people from expressing themselves but rather force them to express themselves." For Deleuze, the dialectical relationship of self-representation and oppression is based in our being suffocated and "plagued by . . . pointless statements."[48] In this sense, Picard's and Deleuze's statements could also be mobilized for a critique of social networks. The patient listening (to the other and to oneself), from within which speech is meant to be composed, vanishes in a culture that exists under the primacy of participation and that scarcely still even perceives the sender and the receiver as separate positions. In such a culture, the public inevitably shrinks back when confronted by sentences like: "There is also more silence in one person than can be used in a single human life."[49] Certain perspectives also seem hopelessly outdated, for example the praise of solitude that Picard's contemporary Saint-Exupéry pronounced: "In the gloomy light of a rainy day, I see, in some small silent town, an invalid, shut away from the world, meditatively gazing out the window. Who is she? What has someone done to her? I, for my part, judge the civilization of a small town by the density of this presence."[50]

When these sentences were written, images that many people today would more likely associate with wilderness inspired hopes of a more intense experience of presence, an intensity that would result not from a permanent flow of information but from the balance between speech and silence, experience and reflection. When Saint-Exupéry, Picard, and many of their contemporaries identified immobility as a criterion for being human, they were issuing an exhortation to interiority that was implicit in a critique of culture, which had been challenging bourgeois busyness since the late eighteenth century. Friedrich Schleiermacher's complaint about his contemporaries from 1800 may serve as an example: "People shy away from looking into themselves, and many tremble like menials when they can no longer avoid the question what they have done, what they have become, who they are. Such business is anxiety for them, and the outcome uncertain."[51]

Restlessness generated by fear has its counterpart in restlessness based on conviction. Germans may think immediately of Goethe: "If ever I plead with the passing moment, / Linger awhile, oh how lovely you are! / Then shut me up in close confinement, / I'll gladly breathe the air no more."[52] This is Goethe's version of Faust's pact with the devil—not a pact that promises to exchange the soul for twenty-four years of youth, riches, and pleasure but a wager that can be won if the individual strives and works hard (*"strebend sich bemüht"*). Goethe takes the self-confident ego from his earlier "Prometheus" poem and inserts the devil in the place of God, for just as Prometheus, challenging Zeus, declares the world to be the result of *his* creative power, so Faust doubts that Mephisto has the means to make him, the restless adventurer and entrepreneur, sufficiently happy to want to halt time. Both works are about the human ambition to shape the world, whereby Faust's drive to act—"Live, life demands, if only for a moment"—already shows Prometheus's Storm and Stress mutating into bourgeois competency and efficiency.[53]

Basically, Faust manages without either God or the devil. He represents the shift to a teleological model of life in which humankind no longer feels itself part of an eternal divine order but sees itself, instead, as an actor in social development. The Enlightenment speaks of the education of the human race, to which every individual must contribute.[54] Improving the world starts with the individual, who, in this sense, even outside the context of religious meaning, does not need to feel like a lost mortal but can conceive his role as part of a larger historical context. Goethe's Faust is a child of the Enlightenment who has long since arrived at nineteenth-century pragmatism. He embodies the most radical reaction to Pascal's *horror vacui*: A defiant activist, he fills the "vacant" space with a social project that can do without God.

This new Faust first surpasses the old one, who has lived on in the heroes of novels like Ludwig Tieck's *William Lovell* (1795). There, too, the reader encounters the complaint about the senselessness and lack of prospects of a life that no longer feels existentially bound to divine temporality: "Is the world not a great prison, in which we all sit like miserable criminals and anxiously await our death sentence?" Here, too, the answer lies in avoiding the moment that lingers. But the life of activity that Lovell opposes to the life of contemplation contains no individual project, much less a social one; instead, it exhausts itself in episodes of indulgence. "Oh, how lucky are those reprobates who can forget themselves and their fate in cards or wine, in a wench or some tedious book!"[55]

As a response to the existential problem of modern man, Lovell remains an exception and cautionary example until the dawn of the "experience society" of the late twentieth century. He becomes a robber and a beggar and ultimately courts death in a duel. Thus, he no longer achieves what for Augustine or Buddha was the turning point from a meaningless life of pleasure and excess to a life lived in the fullness of time. But for Tieck, this was precisely the point; he wanted to valorize the moment in time as the most intensive aspect of self-experiencing, rather than of self-enjoyment. The social utopias of the nineteenth century took a different path, offering models for integrating the individual into the context of broader events that were, in Goethe's sense, nonreligious. What the Enlightenment called progress was now (sometimes in opposition to the actual "progress") envisioned as a social utopia, whose realization is once again comparable to the fulfilled moment in Goethe's *Faust*.[56]

Part of Faust's pragmatism, disguised in Goethe's play as the initiative of Mephisto, is the killing of Baucis and Philemon, the mythical lovers who are not prepared to sacrifice their little plot of land for Faust's construction project. This death prefigures the millions who in the twentieth century would later "stand in the way" as social utopias, in the form of Soviet and Chinese socialism, became realpolitik. The fictional and real dead who lined the waysides of the social utopias bear witness against their potential as alternative models for the construction of meaning, and simultaneously they weaken the moral superiority of the new Faust in comparison to the old one. To this must be added, starting with the awakening of climate awareness at the end of the twentieth century, a profound skepticism toward the Faustian (and Anthropocene) paradigm of a *homo faber* who changes the world in ways that conform to his own image and who, in the process, as is becoming increasingly obvious, destroys the foundation of his own natural life.[57]

This digression into the past is more closely connected to Facebook society than it may initially appear. For when social-political alternatives have discredited themselves, and when religious projects also no longer provide a solution, the question remains: What is life for, and to what extent, if there is no satisfying answer, does communication on social networks promise salvation? If the life instinct, when Goethe's *Faust* and the Enlightenment's idea of perfectibility both appear to be done for, dissolves into mere episodes of pleasure? Does Facebook, with its offer of delegation, help channel the pressure of experience?

The second half of the twentieth century calls the unified concept of a history that could be termed progressive into question also in linguistic and discourse-theoretical terms. Insight into the relativity of the dominant value system leads to critique of the "grand narratives," and the Enlightenment ideal of humanity is criticized as Eurocentric. When, simultaneously, in light of the failure of "real socialism," the end of history is declared, this still occurs under the sign of the modern, Hegelian belief in progress: History ends not in a political standoff, as it seemed to do in the context of postmodernity, but with the free-market system as the sole, unchallenged victor in the battle of the social systems. But the effect is nonetheless similar. Even in the flush of victory, human beings lose the projects that have given their lives meaning and made the future seem more important than the present. The success stories that are told in each case—the end of the grand narratives as emancipation from a one-sided view of reality, or the end of history as the replacement of relativism by a universal social model—are basically a declaration of loss: the loss of a life project toward which the elements of existence could be oriented.[58]

This declaration of loss caps a more long-term historical framing of contemporary centrism that sees the lack of significant and meaningful projects not phylogenetically, as the absence of a social utopia, but ontogenetically, as if it is individuals' life plans that have gone missing. The modern individual, to borrow an analogy suggested by Zygmunt Bauman, is no longer a ferryman who follows the course of the riverbed as he finds it but a seafarer who must seek his own path. If the life of a premodern individual was largely predetermined by his social context, the modern ego is self-created. The result is not independent of the social context, nor is it, any longer, simply determined by it without any resistance. The new direction appeared in the guise of liberation from traditional structures, encrusted role models, and inflexible rules. But while the life of the modern individual was still oriented toward this liberation as its goal, the postmodern person lives entirely in the present, freed from imposed life structures and obligatory rites of transition but also lacking a coherent life narrative into which experiences could be meaningfully fitted. On the one hand, according to Bauman, the postmodern subject wants not to control the future but rather to prevent it from falling into the hands of others. On the other hand, he also wants to prevent the past from having an influence on the present. Thus he cuts the present off at both ends and views time as nothing but an assemblage of random moments: a continuous present.[59]

The next metaphor for humans who have been disconnected from the future is, consequently, no longer the traveler but the hunter. Bauman borrows it from Pascal's hare-hunting analogy but displaces the rationale when he presents the hunt for experiences as being driven not by the individual's fear of being alone with himself but by his fear of social isolation.[60] Here, Bauman follows the currently dominant explanatory model for the publication of intimate details on social networks, which now, since "experience your life" has been supplemented by the additional motto "narrate yourself," is as indispensable as hunting. Within the current theoretical frame and in the practical context of attention economy, however, an important rationale is being jettisoned if the fear of solitude is covered up by the fear of being alone. The desire for recognition and acknowledgment, which explains today's social networks, played no role in the flight into distraction for Pascal and the Romantics—or, as we shall see, for Friedrich Nietzsche and Siegfried Kracauer. Publishing personal experiences may be quite helpful in enabling this flight, and in the meantime certainly has also become a goal of its own, as we look at reality with a "Facebook eye" that seeks to determine how what we have "experienced" can be presented most favorably and with the promise of the most likes.[61] But anyone who overlooks the *horror vacui* behind the need for social recognition is ultimately turning causality on its head and mistakenly taking social networks as the *basis* for obsessive photographing and communicating, while in actuality they merely offer the perfect opportunity for the hunter's flight from melancholy. It is not, as is often claimed, the social network that separates us from social life; it is the felt lack of a real life that makes the social network so attractive as a "respectable" way out.

Bauman awards utopia status to "hunting" society because, like all utopias, it puts an end to human suffering. It does this not by answering the question of the meaning of life but by abolishing the question itself in favor of an endless series of individual lived experiences. Beyond any broader ambition to improve the world, the individual finds happiness in an unending series of nonreflective life events. The gist of this historical and philosophical perspective only comes across as truly ironic if, as Bauman does, we continue to believe in the categorical imperative of self- and world improvement.[62] If, on the other hand, the life elixir of postmodern man is hedonism, pleasure in the here and now, then Tieck's Lovell and Goethe's Faust meet in the present in a paradoxical embrace, as both desire for and flight from the moment in time. Life in the moment is unbearable precisely because beyond the moment there is nothing for the lived experience to anchor itself to or to aim toward, nothing that might give it weight. Tarrying in the moment—this is Mephisto's revenge—contains a profound vacuum, even in its most fast-paced form. The moments can be borne only when they are in movement; otherwise we cannot avoid the question of what we are doing and what we are, or even the admonition "You must change your life"—presumably meaning something more than self-optimization in the fitness cult. It is not only the monuments of culture or nature that create a quandary for us but also the moments we spend with ourselves, not only the reflective quiet of a rainy day that can become dangerous for us but also the vacuum we recognize in moments of pleasure. In Facebook society, the *horror vacui* is expressed as the shock of plenty.

It is possible to explain Facebook's psychological relevance for late- or postmodern humanity in various ways. Faust is one variant; Bloom is another. In the latter case, the philosophical reference point is not Goethe and Tieck but Nietzsche and Heidegger. In the Tiqqun collective's *Theory of Bloom*, Leopold Bloom, the main character in James Joyce's *Ulysses*, is the "last man" and "rootless man," the man of "non-participation" and "non-belonging."[63] The "fundamental tonality of being" embodied in Bloom is to be found in the subject's "withdrawal from the world, and vice versa." No longer wedded to earthly or heavenly goals, man "cannot take part in the world as an inner experience." Unlike the Romantics, he also lacks "the recourse of an interior desertion": "All attachments are replaced by that of surviving." As with Lovell, it is the flight into distraction and spectacle, as a kind of "existential tourism," that promises salvation. "The spectacle has relieved Bloom of the burdensome obligation to be."[64]

Bloom's salvation lies in his capacity to become more and more intoxicated by less and less. Against this withdrawal into the society of spectacle, Tiqqun calls for revolt, with a rhetoric that feels as purely justified as Saint-Just in the German playwright Georg Büchner's play *Danton's Death* (1835), when Saint-Just defends Robespierre's revolutionary terror with high-flown words about the mission of history. Anyone who takes into account the losses of the French revolution and the following revolutions—all made in the name of improving the world—will be skeptical of future revolts and perhaps even speak of the "disaster of the promise of emancipation" that "[wakes] us from a sleep-filled life of consumption only to throw us headlong into fatal utopias of totalitarianism."[65] Herein lies the absurdity and aporia of the present: Consumer culture signifies not only distraction from the responsibility of being but also, and at the same time, liberation from cultural, national, and ideological references as differences. As Tiqqun notes, "And this Common resulting from the estrangement of the Common, and formed by it, is nothing other than the true Common, unique to men, their originary alienation: finitude, solitude, exposure. There, the most intimate merges with the most general, and the most 'private' is the most widely shared."[66]

From a perspective less rebellious than the one expressed by Tiqqun, this communality of loneliness, beyond shared values, is not the problem but the solution. For only at the zero point of a connection with reliable beliefs are human beings so reduced to the most human attributes— "finitude, solitude," in Tiqqun's formulation—that a community beyond shared points of view, and hence also beyond the exclusion of "others," becomes thinkable: the community of communication. In a formulation by the French philosopher Jean-Luc Nancy, whose concept of community without commonality will be of interest again in the following, it is a community "that refuses to set to work on its own establishment, and thus preserves the essence of an endless communication."[67]

At the time when the Bloom text was written, it was still too early to reference social networks, and when the authors later turn their attention to the digital media they give their cultural critique a political turn, in order to investigate the connection between cybernetics, control, and revolt. If we apply *Theory of Bloom* to the current era of social networks, we find that escape is now sought less in intensified experience, or as "existential tourism," than in the *communication* of the experiences. The social networks generate a society precisely out of individuals' alienation from all reliably meaningful bonds. Tiqqun's philosophy has no place for this type of approach, particularly not when the "salvation-bringing" medium is as vehemently committed to consumer culture as Facebook. Other people, of course, see this differently and treat the social networks, either hesitantly or with enthusiasm, as a new model of community.

For these observers, "individuals, woven into and lost in the techno-social network," are understood as a community whose activity consists in nothing other than "the longing for the opposite number of one's own existence and action," for "the thou, the recognition by his gaze, even if this gaze, in its concrete realization, may be imagined." The perhaps erroneous impression can be ignored as long as the feeling of community actually occurs; the decisive thing is the *feeling* of community based on a common practice of media usage rather than on shared opinions.

Even if the individuals who are linked by a social network do not constitute "a *societas* in any traditional sense of a binding communication community," the "connected individuals"—so the argument goes—"cannot possibly be completely random singularities, because with all its attractive user surfaces and endorsements the telematics *dispositif* has long since had an identity-creating function for them. Through their techno-communicative activity with each other they constitute a community to which they have a positive relationship; just how they want it."[68]

If the telematics *dispositif* has formed the basis of the identity, then a feeling of community is being generated not by shared interests but by a shared medium. This difference can be understood with the help of the "imagined community" described by Benedict Anderson, which is created not primarily through its shared contents but by the shared use of the same language. It is the medium, not its contents, that creates the community. The members of the community communicate with it in a dual sense: as a means but also as its addressees. The social network Facebook is the (imagined) community to which we feel we belong when we communicate with the (real) community of our Facebook friends. Facebook is the "language" that creates community.

The assumption of a new model of community qua medium is also in evidence when Facebook is celebrated as "an attempt at liberation from the mechanics of purposefulness," as "training for *désinvolture*," and as a utopian space that offers and permits new forms of rapprochement and makes it possible, "in view of the complexity of social offerings, to remain lovers of the moment."[69] In this case, the "desire for a non-purposive language of the moment" cures Faust's opposition to lingering with a dose of the "presentist fervor of Twitter and Facebook." The result is speaking for the sake of speaking, commonly known as "small talk" or, in academic terminology, "phatic communication": communication that aims at nothing more than the act of communicating. This mode of communication does not require prior consensus, nor does it seek to create it; instead, it simply insinuates it as an aspect of the communicative gesture. Communication no longer serves the purpose of an exchange about contents but aims solely at the creation of an external connection.[70]

Is Facebook, which so mercilessly defiles every individual action with advertising, suitable to serve as the bearer of social hope because it succeeds in creating a community without consensus? Is there true life on the false network? The answer ultimately touches on questions of principle that lie beyond the phenomenology of the digital media. What is it possible to hope for, if the postmodern end of the project of modernity leaves only commerce as a system of relations and if the return to ideology or religion is marked as a failure in historical-philosophical terms? Does the ruse of reason lie in technology? Could it be possible that on a social network like Facebook—beyond and despite its incriminating business model—a community is forming that is as noncommercially oriented and nonutilitarian as is being suggested here? Can Facebook be thought as the technical equivalent of the community of the alienated that Tiqqun critically identifies? In this case, in contrast to the view presented by Tiqqun, we should no longer think of the sharing of the most private things philosophically, as a common thrownness and lostness, but sociologically, as activity on the social network: as (willed) communication of the private. The "communism" of this sharing is grounded in lack, a lack common to all: finitude, loneliness, thrownness.

It is easy enough to assume positions that run counter to this starting point, particularly after the "critical turn" in internet studies that has turned the previous, disappointed enthusiasm about the potential of the internet into acerbic critique. The obvious bases for this critique are rehearsed often and extensively and may certainly not be ignored. But at this point in our reflection they must not obscure the question whether, ironically, it is not precisely the capitalization of private life that ultimately overcomes the isolation that has accompanied the all-embracing commodification of postmodern society. The attempt to open a narrow line of sight that looks beyond the usual view of Facebook as a site of commercialization and producer of authoritarian knowledge must pursue this unorthodox thesis with corresponding perseverance.

EXPERIENCE VANISHING IN THE MAELSTROM OF EXPERIENCES

One of Fernando Pessoa's imagined figures, or heteronyms, Alberto Caeiro, is a mystic of intentional nonknowing, a worshipper of things as they are, who sees in a flower nothing more than the flower and in the sun no more than the sun. Caeiro does not interpret the world; he does not shape it into metaphors, much less into a system of structured thought. He asks questions about nothing and gives answers to no questions. He seeks no more profound meaning, and this, without a doubt, represents the more profound sense of his existence. Caeiro embodies the antimetaphysical poet who, unlike his brooding creator, Professor Pessoa, is happy.[71] This is of interest here insofar as Caeiro's stance seems to correspond to the more affective than reflective link to the present that obtains on many social networks. Moreover, it reminds us of various theoretical initiatives that, in recent decades, have propagated an aesthetic of presence and the performative that transcends any attempt to grasp reality aimed at sense and meaning. Here, it is helpful to have a look at historical precedents and current parallel developments.

Hans Ulrich Gumbrecht, in opposing the Cartesian principle of interpretation, or "meaning culture," argues for the concept of "presence culture," which aims at "moments of intensity" that may contain "nothing edifying . . . , no message, nothing that we could really learn from them," but that on the other hand are closer to, "*more in sync with the things of this world.*"[72] This affirmative turn toward the present expressly counters the imperative of the Enlightenment and the "so-called 'generation of '68,' with its by now grotesque fixation on an exclusively 'critical' worldview." Presence culture—Gumbrecht refers repeatedly to this political background—represents a release from the permanent duty to be in movement "both in the sense of the never-ending 'historical' changes imposed upon us, on all different levels of existence, and in that of the self-imposed obligation that makes us want constantly to 'surpass' and transform ourselves."[73]

Gumbrecht's "presence culture" is the philosophical antithesis of critical theory, which counseled mistrust in "all letting oneself go, for it includes pliancy toward the superior might of the existent."[74] Presence culture replaces the notion of improving the world with that of embracing the world. The cultivation of intensive moments of thought-free feelings vanquishes the problem of human mortality in a unique fashion: not by participation in deeds that help shape the world but by the experience of oneness. Is Gumbrecht the secret mouthpiece of the Facebook generation? His plea for the meaning-free passion of sheer happening is undoubtedly very close to the absolute embrace of Caeiro. Does it also provide theoretical cover for phatic communication on Facebook? Could Gumbrecht's historically and philosophically determined concept of presence culture be evidence of the fact that phatic communication on Facebook was not (solely) a result of technical development but corresponds to an already existing, already theoretically expressed need? To answer this question, we should take a look at the golden 1920s, when experience in the strong sense, as gained over time and linked to understanding, was watered down into individual lived experiences. Not by chance, Gumbrecht himself provides the point of entry for this review.

With his book titled *In 1926: Living at the Edge of Time*, which appeared in 1997, Gumbrecht offered an example of historiography in the mode of the culture of presence: a report on cultural, political, and academic events of the year 1926, in the form of a dictionary. An "original treatment of the problem of representing history after the end of the grand narrative," the American historian Hayden White calls it on the back cover. White's comments on the development of historical writing will occupy us again in chapter 2, which deals with the transition, in the late eighteenth century, from the accumulation of factual material to the creation of coherent narratives. Here, looking at 1926 and *1926*, we will examine the contrast between the two forms of experience.

In the entry "Reporters," Gumbrecht distinguishes lived experience (*Erleben*) from perception, on the one hand, and from substantial, longer-term experience (*Erfahrung*), on the other. "*Erleben* is situated between 'perception' and 'experience.' It adds to perception by focusing on what is being perceived, but it does not include interpretation. *Erleben* is more than just sensory contact with the environment—but less than the transformation of the closely regarded environment into concepts."[75] The place where the shift from the mode of *Erfahrung* to the mode of *Erleben* occurs is newspaper reporting, whose goal of objectivity is considered to require speed in writing and a merely superficial contact with the objects reported on. "Only speed can prevent the reporter from getting caught in the depths of interpretation and experience," notes Gumbrecht. "Insofar as the reporter refrains from all interpretations and judgments, simply 'carrying back' impressions from his direct contact with the world, he is 'objective.'"[76] This calculation (superficiality +

speed = objectivity) will come up again in the discussion of status updates on Facebook.

It is not surprising that "raging reporter" Egon Erwin Kisch
serves as the central point of reference for Gumbrecht's entry
on the subject and is also the source of his catchwords. Kisch
recommends compensating for the metaphysical lack of orientation that characterizes the present by intensive proximity to
its phenomena: "Those who succeed in clinging to reality manage to survive the disappearance of ideas and values." In Gumbrecht's words, "The restless life of the reporter and his surface
view of the world are linked with the collective—and often
repressed—fear that ultimate truths are no longer available."[77]

This focus on the external became an adequate way to relate
to reality at a time when war and the dominance of technology
seemed to have rendered depth-psychological interpretations
and emphatic description inappropriate. Along with the "raging
reporter" as the symbol of a society that was already sped up and
losing its orientation, the "cool persona" now also became the
dominant type of perception, embodying a consumer type that
registers its surroundings with distance and without emotional
engagement.[78] It is the type of the "New Objectivity," a trend
in art, literature, photography, and architecture that shaped the
aesthetics of the Weimar Republic, using a language "with no
lyrical fat": "hard, tough, trained . . . comparable to the body of
a boxer."[79] It is no surprise that photography was invoked here
as the genuine medium of this "impassiveness," because, as they
said, it registers things "outside the sphere of the emotional."[80]

Photography is the "cold" medium of information capture.
As Brecht noted, it is "the possibility of a *re*production that
masks the context." As Jean Baudrillard would later remark, it
represents the "anti-philosophy of the object," which "reports
on the state of the world in our absence."[81] For Baudrillard, the
drama of photography consisted in the "struggle between the will
of the subject to impose an order, a point of view, and the will of

the object to impose itself in its discontinuity and momentary quality." What is interesting about his by no means original perspective is the external resolution of the conflict, when Baudrillard understands the use of the camera as a reaction to a specific consciousness of reality: "Perhaps the desire to photograph comes from this observation: Looked at from a general perspective, a perspective based on meaning, the world is quite disappointing. Seen in detail and caught by surprise, as it were, it is always perfectly evident."[82]

Evidence instead of transcendence—this also works for moving images, as a contemporary of Kisch and Ernst Jünger observed. Béla Balász's *The Spirit of Film* (1930) contains a section called "Flight from the Story." The passage reveals much about the relation of the camera to cold observation (or the "clear gaze"), using the example of Ernest Henry Shackleton's Imperial Trans-Antarctic Expedition (1914–1917), during which the ship became stuck in the ice and ultimately broke apart.[83]

> This is a new form of human consciousness, which is given to man by the camera. As long as these men don't lose consciousness, their eye looks through the camera's lens and imagines the scene in the lens as present. Presence of mind becomes presence of the image. —Shackleton's last hope, his ship, is breaking apart under the pressure of massive ice floes . . . he shoots . . . they drift onto the iceberg, and the iceberg is melting under their feet. . . . This is shot. . . . For the camera provides security. It is a form of self-consciousness. The inner process of taking account of something has been turned inside out. The "clear gaze" is mechanically fixed so it can be held longer. The self-control of consciousness formerly existed as a series of internal images; now it is loaded into the camera as film, where it functions mechanically and is also visible to others. . . . We don't keep shooting as long as we are conscious, we are conscious as long as we keep shooting.

This film-shooting consciousness is symptomatic for the new model of experience—as both *Erfahrung* and *Erlebnis*. The camera becomes a site of externalized reflection; consciousness is reduced to the act of witnessing, which provides a way to hold on to what is being lost at the very moment when it is disappearing. Another, less life-threatening but still dramatic threat of destruction looms in the form of the loss of orientation in the accelerated present of the twenty-first century. The externalization of reflection as witnessing that was observed in 1930 continues in the form of sharing on Facebook, Instagram, Snapchat, etc. When Balász writes "presence of mind becomes presence of the image," we should ask, employing Agamben's concept of contemporaneousness through distance and reflection, whether precisely this presence of images reduces our *understanding* of the present. To what extent is the "inner process of taking account," in Balász's terms, turned outward, today as well, when everyone is the raging reporter of his own life? Don't processes of automation in social networks and through apps extend the "cold gaze" even into such personal areas as autobiography? Before we discuss this process, a few words should be said about the social psychology that formed the basis of the paradigmatic shift that took place in the 1920s.

Walter Benjamin also remarked on the sealing off of information from experience in contemporary journalism, although his evaluation was distinctly different from that of Gumbrecht or Kisch. "If it were the intention of the press to have the reader assimilate the information it supplies as part of his own experience, it would not achieve its purpose. But its intention is just the opposite, and it is achieved: to isolate events from the realm in which they could affect the experience [*Erfahrung*] of the reader." Benjamin distinguishes between "information" and "sensation," on one hand, and "stories," on the other, stories being the earlier form of communication in which the subjective and the personal were still present. "A story does not aim to convey an event per se, which is the purpose of information; rather, it embeds the event in the life of the storyteller in order to pass it on as experience to those listening."[84]

The difference between story and information parallels the distinction, which is constitutive for Benjamin's thought, between *Erfahrung* and *Erlebnis*. In the *Arcades Project*, he says: "Experience [*Erfahrung*] is the outcome of work; immediate experience [*Erlebnis*] is the phantasmagoria of the idler."[85] For Benjamin, unlike the Romantics, work and leisure are related to the mental activity of working through lived experience, which is required in any case. Here, idleness is not positively defined as contemplation but negatively, as distraction. Benjamin assumes a diametrical opposition: *Erfahrung* is authentic and rich in consequences, for it reaches into the present as wisdom that derives from the past; *Erlebnis*, on the other hand, is superficial and without consequences; as intensified perception, it remains reduced to the present moment. In the short text "Experience and Poverty," to which both Agamben and Tiqqun refer, Benjamin confirms something that, after the given description of mass society, is hardly surprising: "Experience [*Erfahrung*] has fallen in value." What *is* surprising is what Benjamin says next:

> Poverty of experience. This should not be understood to mean that people are yearning for new experience. No, they long to free themselves from experience; they long for a world in which they can make such pure and decided use of their poverty—their outer poverty, and ultimately also their inner poverty—that it will lead to something respectable. Nor are they ignorant or inexperienced. They have "devoured" everything, both "culture and people," and they have had such a surfeit that it has exhausted them. No one feels more caught out than they by Scheerbart's words: "You are all so tired, just because you have failed to concentrate your thoughts on a simple but ambitious plan."[86]

There is not too little experience, but too much of the kind of experience that makes people poor in hope. This experience is not only a consequence of the trauma of the First World War and the following "total absence of illusion about the age."[87] According to a very early text of Benjamin's, it is also related to the disillusionment an adult feels following the promising period of youth. Youth is followed by the "grand 'experience' [*Erfahrung*], the years of compromise, impoverishment of ideas, and lack of energy. Such is life. That is what adults tell us, and that is what they experienced [*erfuhren*]." For the twenty-one-year-old Benjamin, in his aggressive response to "Philistines," the result of this disappointment is the turn to an experience (*Erleben*) that is "spiritless."

The writer of "Experience and Poverty," at forty-one, has become more forgiving toward his fellow citizens who have grown tired of experience, although—or because—he has meanwhile raised the ontogenetic problem to the level of phylogenesis. Apathy and hopelessness are no longer the result of a midlife crisis but a product of the social status quo. Even before his late theses "On the Concept of History," Benjamin says: "The concept of progress must be grounded in the idea of catastrophe. That things 'go on' *is* the catastrophe. The catastrophe is not what is expected to happen next, in each case, but what in each case is [already] given."[88] Benjamin's disappointment is palpable. Fatigue and the lack of a "grand plan" are understandable in light of social development. Benjamin responds to this disappointment with the messianic responsibility to recognize the hopes and claims of the past on the future and to fulfill them. "The past," he writes in his second "thesis" on the concept of history, "bears with it a secret index by which it is consigned to redemption." The means of recognizing the signs of this secret index is the "dialectical image," which Benjamin, in his epistemology, presents as an empathic solidarity with the secret side of reality and as an alternative to the objective, photographic model represented by the reporter and "cool persona."[89] In this context, Benjamin's contrast of *Erfahrung* with mere *Erlebnis* assumes historical-philosophical dimensions and firmly retains what Gumbrecht no longer wants to take responsibility for— the "trans-epochal context of validity and responsibility" of the contemporary era vis-à-vis the past, toward the creation of a better future.[90]

It is worth nothing that both Agamben's and Tiqqun's diagnoses of the present place a central emphasis on the shrinking of experience (*Erfahrung*) that Benjamin described in 1933, in his essay "Experience and Poverty." With this, a historical frame is established for Bloom's role as a central social figure: James Joyce's *Ulysses* appeared in 1922, Heidegger's *Being and Time*—with which Bloom becomes "the central non-subject of philosophy"—appeared in 1927. Bloom, according to Tiqqun, becomes a mass phenomenon at the moment of the "exhaustion of metaphysics."[91] It is also the very historical moment when cinemas and country inns become sanctuaries for the mass of employees, who, as Kracauer describes the era, are "spiritually homeless . . . without a doctrine they can look up at or a goal they might ascertain."[92] The postmodern experience society of the end of the twentieth century carries this burden of homelessness and lack of orientation with it into the new millennium, in which mobile and social media convert the model of experience into a structure of sharing and interaction. Facebook society should also be seen as a contemporary attempt by mass society to overcome its metaphysical homelessness through participation culture.

Benjamin offers up a link between the two historical phenomena—one at the beginning of the twentieth and the other at the beginning of the twenty-first century—with his claims about media theory, which, in the context of his historico-philosophical and epistemological reflections, allow us to draw a direct link to Facebook, Instagram, Snapchat, and other sites of self-representation. Thus, only a few pages after the entry on progress as catastrophe, he writes: "The souvenir is the complement to 'isolated experience' [*Erlebnis*]. In it is precipitated the increasing self-estrangement of human beings, whose past is inventoried as dead effects." Souvenirs, for Benjamin, are "secularized relics" with a new point of reference: "The souvenir comes from the defunct experience [*Erfahrung*] which thinks of itself, euphemistically, as lived experience [*Erlebnis*]."[93] The souvenir provides a socially acceptable form of liberation from experience, through which people may "make such pure and decided use of their poverty—their outer poverty, and ultimately also their inner poverty—that it will lead to something respectable."[94]

With the benefit of historical distance from this conclusion, we can add: Photography is respectability enhanced by an individual's own activity, which, when directly communicated on social networks, elevates that action to interaction. It is, however, rare for the shared photo to constitute an actual exchange of experience (*Erfahrung*). Sharing, liking, and commenting on "souvenirs" is ultimately not much more than a brief halt in the presence of the messages, a lingering in the state of communicating. This permanent flow of short-term experiences (*Erleben*), given the quick pace of messaging, does not permit the very thing that holding still promises: cohesive, cumulative experience (*Erfahrung*).

The natural reflex in response to the acceleration of society is the call to stop moving, to switch from *vita activa* to *vita contemplativa*.[95] Gumbrecht's plea on behalf of embracing the world and for experience that abstains from interpretation can be understood as a version of this kind of awareness. His posthermeneutic concept of presence culture may contradict both Goethe's will to shape the world and Benjamin's responsibility toward history, but in replacing deep thinking, as the central characteristic of the culture of meaning, with deep feeling, in the sense of passion, abandon, and the removal of boundaries, it also differs from the phatic communication that is practiced on social media. Despite the evident proximity to the phatic culture of networks, the difference is impossible to overlook. Gumbrecht is closer to the absolute embrace in Caeiro than to the group snuggling on Facebook, which may have no object at all. His point of departure rebels against traditional models of ascribing meaning but is actually quite conservative in comparison with the standards of the social networks.

Facebook, Instagram, Twitter, WhatsApp, and Snapchat celebrate the self, with a mixture of permanence and immediacy that goes beyond Caeiro's profundity and obsession with objects or, as another example, Walt Whitman's powerfully worded self-articulations. While the opening line of Whitman's famous "Song of Myself" ("I celebrate myself") enacts its self-justification in the performative energy of its self- and world description, on social networks self-representation is mostly delegated to the mechanism of the camera and rooted in a profound lack of self-initiative. The result of this development is a network autobiography that emerges automatically and posthumanly, simultaneously bypassing its object/subject, the act of reflection, and meaningful experience altogether.

2

AUTOMATIC AUTOBIOGRAPHY

Photography grasps what is given as a spatial (or temporal) continuum; memory-images retain what is given only insofar as it has significance.

—Siegfried Kracauer, "Photography," 1927

Dear diary, it's only now that I get around to writing to you again because all week I was reading a book that was so fascinating I didn't find time to write anything down. It was Hermann Hesse's *Steppenwolf*, which Christian recommended. Christian said if I want to know more about breaking free of boring, petit-bourgeois life I should read this book. He was right! The book is full of sentences I'd like to hurl at so many people! For example this one: "In reality, however, no I, even the most naïve one, is a unity, but instead it is an extremely diverse world, a little heaven full of stars, a chaos of forms, levels and states, of inheritances and possibilities." I think this is exactly the way it is, because . . .

Except for the salutation, everything in this diary excerpt is authentic. It was Christian who recommended the book twenty or thirty or forty years ago, and reading it actually did keep the author from writing in his diary for a whole week, resulting in an entry that was that much longer—ten pages or more, enough to keep him busy for a whole evening. There are some crossed-out passages and a lot of exclamation points, and afterward the writer was able to quote the mentioned passages repeatedly over the years, if not always word for word. Those were the days, when along with the things people had experienced they also wrote down their feelings and ideas! When things that had happened were endowed with meaning at the special moment when they were being written down—thoughts that came to completion in the act of writing. Insights that helped a person grow. Does anyone, anymore, on a Sunday evening, try to reflect on the week that has just passed? Are there still people who keep diaries?

If you are conscious of the fact that diaries, unlike poetry albums, were never a mainstream phenomenon, and if you think you can count words as you do numbers, you might conclude that the twenty-first century is the dawn of a new era of self-knowledge, for never has there been so much writing as there is now.[1] But if you judge the situation less quantitatively, you will also want to know what kinds of words are being uttered, in what circumstances, and what form they take. In principle, it is clear that when you are writing on social media like Facebook you express things differently than you would in a diary and that a hundred status updates don't add up to a deeply felt and thoughtful comment. The more the communication form formerly expressed in the diary is occupied (that is, displaced or redefined) by social networks, the less we can look forward to the kind of self-reflection the diary can offer. When autographical writing is hailed as the "master form of the 21st century,"[2] we should therefore ask how self-representation and self-knowledge are related to each another in the context of the technical and social frameworks that now shape our social communication.

A first answer to this question is that social networks are not only used for conscious self-branding; they also prompt their users to engage in unconscious self-revelation. The self-revelation occurs more implicitly than explicitly, depending on the network—naturally, it is more carefully designed on a network for business contacts like LinkedIn than on Facebook. Showing replaces saying (or writing), as photos bear witness to more or less spontaneously experienced moments, and likes more or less spontaneously express personal preferences.[3] Even textual updates and comments, when they happen as part of spontaneous communication, are no guarantee that the act of making something visible to others also promotes self-knowledge. Yes, precisely those comments that are less controlled can be harbingers of potential self-knowledge because they lack any subjective communicative intent and are therefore free of unconsciously imposed rules for the construct of the self. Users are free, after some time has passed, to sift through the photos, likes, and comments on their pages in search of themselves. The question, however, is: How many people take the time for this self-encounter, and does it go beyond mere wall cleaning and the removal of old postings that have become embarrassing?

Along with spontaneous showing, there is also an automatic showing, which takes place when the mechanism that inserts specific actions by internet users—visits, likes, shares, or comments—directly into their timeline is activated.[4] When this happens, individuals no longer describe themselves more or less implicitly, through their actions, but instead it is the actions that describe the individuals. The subject's "internal automatism" is replaced by the external automatism of the system the subject has become a part of. A popular example of this automatism is self-tracking, which promises self-knowledge through numbers instead of words. Here, the self's body produces data that is independent of the person's own consciousness (even though she initially agreed to it) and that makes statements about the person independently of her self-understanding and the values that have been internalized in her self-construct. This is generally true even when the data must be entered by hand into the relevant app. For as long as it is not a matter of describing one's own mood but of factual indicators (physical movement, foodstuffs consumed, sleep behaviors), there is little incentive for reflection.

Facebook's interest in acquiring as much data as possible for statistical computations and its creation of individual profiles in the interest of effective advertising are well known, and while occasional revelations about the extent of secret data mining may be shocking, they don't come as a surprise.[5] The fact that self-representation on social networks like Facebook produces economically and socially exploitable knowledge about the social body is not what is decisive here. More important is the question of what effects the context of this self-representation has on the subject's own self-perception and self-knowledge. Here, it is important also to take account of the outright avant-gardist aesthetics that this context creates: the paradoxical phenomenon of a simultaneously actionist and postactive, automatized autobiography, one more lived than narrated by its subject and "author."

From a media-theoretical perspective, social media, as a new genre of autobiography, mean a shift from writing to photography. This is not because there are more images than text on Facebook but because the reporting takes place in the mode of mechanical reproduction typical of photography. Activities on the internet are reproduced on Facebook exactly as they occur. From a historiographical perspective, storytelling on Facebook turns back to domination by numbers. This suggests that we should turn our attention to a source of history writing for which the simple fact of events was more important than their narrative coherence.

PHOTOGRAPHIC TEXTS

In the beginning was the number. This characterization of the history of historiography would be permissible if we skip over Homer, Herodotus, and Tacitus and start with the medieval annals, which order history not by events but by years. "Hard winter. Duke Gottfried died" is an example from the *Annals of St. Gall*, part of the *Monumenta Germaniae Historica* for the year 709. Two events without any context or narrative setting. And when it states, for the year 710, "Hard year and deficient in crops," there is no explanation of the causes of the poor harvest.[6] What mattered were not events and the relations among them—not at all. Famine and war were not the actual event but merely witnesses to the passage of time in "the fulnes [*sic*] of the 'years of the Lord.'" The passage of time was news enough, for it enacted the connection between the beginning (the birth of Christ) and the end (the Last Judgment). Time was thought cosmologically; the years, in their numerological uniqueness, were its phenomenological element. And because time came from God, the latter was the real hero of the events. This, not war and hunger, was what history was meant to proclaim. Thus it is quite natural when the *Annals of St. Gall* also include the years in which nothing happened and finally conclude with an abbreviated listing: "1057. 1058. 1059. 1060. 1061. 1062. 1063. 1064. 1065. 1066. 1067. 1068. 1069. 1070. 1071. 1072."

From ascribing value to the dates of the years themselves, it was a long way to writing history without naming any years at all. In between came the chronicles, as a method of recording events that were both contemporary and part of a history—of a city, a royal family, a business, an individual. Narrative history, by contrast, offers a finalized, retrospectively recounted and coherent assemblage of events that, as a kind of "historical law of the conservation of energy," are anchored in a solid network of causes and effects. During the eighteenth century, the accumulation of facts without perspective became the object of critique, and the history of isolated events gave way to the universal history of coherently connected happenings, a history that, in the end, was quite logically conceived as a "history without years and names."[7]

Historiography became narrative at the very time—the end of the eighteenth century—when it was proclaiming itself a science by projecting the narrative forward, from the level of reception, where the connection between a hard winter and the failure of the harvest was always potentially available, to the level of production. Chains of causality were now part of the business of historiography. To offer "only what actually happened," the era was convinced, "would be to sacrifice the actual inner truth, well-founded within the causal nexus, for an outward, literal, and seeming truth."[8] Once "inner" truth became the crux, as a way to view the past from the perspective of the present, a paradoxical equation emerged: "The historian is one who prevents history from being *merely* history."[9] This shift in the methodology of writing history raises the question of how dense it is possible for description to be without it becoming narrative and how porous it has to be to avoid the risk of unraveling the chosen narrative. The German historian Johann Christoph Gatterer gave an astonishingly honest answer to this conundrum in his programmatic text from the year 1767, *Vom historischen Plan und der darauf sich gründenden Zusammenfügung der Erzählungen* (On the historical plan and the connection of the narratives based on it): "Events that do not belong to the system are now, for the writer of history, so to speak, non-events."[10]

The disappearance of actual events from the spirit of a narrative, along with a preference for "real" truth in lieu of the "literal" kind, are themes that would return a century later, at a time when both painting and literature were being defended against the new medium of photography. In the mid–nineteenth century, literary realism was judged to have "daguerrotypical similarity" to the representation of reality; people accused it of "idolatry of the raw material." Literature, in effect, can never be as objectivist as photography, and photography, in effect, is less a comment-free representation of reality than the representation of a personal relationship to reality, as expressed in the choice of motif and the moment when it is captured, as well as the perspective used and focal length employed, not to mention the choice of camera and type of film. Nonetheless, painting and photography represent two essentially different methods of producing an image. The distinction is therefore justified and ultimately applies to literature as well: Painters (writers) must decide how to represent the object that, at times, exists only before their inner eye. In other words, they follow their *own* understanding of the "truth" of a thing. In *photo-graphy*, on the other hand—this is the origin of the name—the object writes itself into the photo with light, a situation that inspired the photography pioneer William Henry Fox Talbot to speak of the "pencil of nature" and led the philosopher of semiotics Charles Sanders Peirce to define an indexical type of signs, arguing that a photo is the direct, physical consequence of whatever was in front of the lens, just as smoke is a consequence of fire or a footprint the result of a step.[11]

This physical relationship between sign and signified is also, for the most part, characteristic of Facebook on a grand scale. The written comments that are automatically presented in the news feed and activity log, referring to articles recommended, videos viewed, and music listened to, are indexical from the perspective of semiotics because they are the direct result of the action represented by the sign, not a retrospective description or, at least, an announcement that the action has occurred. They are the "smoke signals" of our online existence. This indexical relationship also basically applies to the texts we compose ourselves, the status updates and comments that appear in our own Facebook Timeline exactly as they were expressed on the homepage. Here, there is no corresponding later entry, as there was in the case of the traditional diary, where it says: "Talked to Christian today about the book and told him that . . ." Now, what we say is already stored at the instant when it is expressed. Facebook's Timeline is a "photography" of events, including communicative acts. In the programmed feedback mechanism of the social network, the event reports itself in real time: it *is* the report.

The technical shift from conscious description to automatic recording means a return from the "real" truth to a "literal" one, in the service of even those events that, as Gatterer noted, do not belong to the system. The indexical nature of the entries prevents the past from being made present again in ways that are emphatic, related to the present, or guided by theory. In 1927, the German essayist and media theorist *avant la lettre* Siegfried Kracauer formulated this specific quality of photography as a loss of meaning: "Photography grasps what is given as a spatial (or temporal) continuum; memory-images retain what is given only insofar as it has significance."[12] The French philosopher Jean Baudrillard would later dramatize Kracauer's declaration of loss by invoking a conflict between the object, as given, and the perceiving subject: "Against the philosophy of the subject, of the gaze, of distance to the world in the interest of comprehending it better, stands the anti-philosophy of the object, the decoupling of objects from each other, the philosophy of the aleatory sequence of partial objects and details."[13] Photography allows objects to prevail, in their discontinuity and momentary quality, independent of the subject's gaze and in principle also in opposition to his perspective and interest. As described by Baudrillard, photography, inasmuch as it eliminates the distance that is necessary to understand the world better, forestalls the contemporaneousness that Agamben invokes.

Facebook—and this is even more true of an app like Snapchat—makes Baudrillard's finding paradoxically more acute. Now, thanks to the sequence of fragmentary experiences and expressions of the Facebook user, all of which are registered and delinked as the here and now of what was once previous, the "philosophy of the subject" is confronted by the "antiphilosophy" of the very same subject. The distanced gaze is lacking, since Facebook, and Snapchat to an even greater extent, permit no distinction between the subject who experiences and the subject who reports. Distance, now, is still possible only in the receptive mode, which corresponds nicely to Zuckerberg's ideal of "frictionless sharing"—of everything, all the time, and above all automatically. The immediacy and nonsubjective character of the report reinforces the moral imperative of authenticity and radical transparency that defines the Facebook community's sharing ethos. Authenticity—in an about-face from earlier historiographic and media-theoretical positions—is viewed as consisting in the automatism of documentation, which is used as a weapon against the "distortion" of retrospective reporting.

The consequence of this departure is a fundamental change in the philosophy of information, from information as an affirmative act to information as the consequence of a quasi-unconscious sharing automatism. The song you hear on Spotify, the film you watch on Netflix, and the article you read online are not communicated to your Facebook friends because you liked them but because you heard, saw, or read them. The communication loses its subjective stamp and hence its value as something that, from the perspective of the sender, is worthy of being communicated. But precisely for this reason, from the perspective of the database behind all the Facebook pages, there is a gain in reliability. As the ex post facto weighing of experience in diary mode is replaced by "insular" reports in real-time mode, even the strategic temporal placement of an update becomes subject to the law of attention economy. The subjective description of the events gives way to their mechanical reproduction.

NARRATIVE IDENTITY

Just as, for Kracauer, it was not the photograph but the remembered image that constituted a person's "actual *history*," so, in the theory of the French philosopher of narrativity Paul Ricœur, time is only human time to the extent that it is "narratively articulated."[14] Humans—this is the basic idea in both cases—must weave the events of their lives into a coherent, meaningful web of relations in order not to feel homeless in them. From this perspective, neither the isolated representation of an event nor the chronological accumulation of many events makes sufficient sense. Moreover, for Kracauer, Ricœur, and likeminded thinkers, the primary addressee of an autobiographical narrative is the narrator herself, not (only) when she later reads what she has written but (above all) in the here and now of the writing process itself. The identity value of the autobiographical act lies less in its documentary than in its performative effect. As the I speaks about itself, it creates itself—this is the core conviction of narrative psychology: "our experience of human affairs comes to take the form of the narratives we use, for we use them not only to tell, but also, and first of all, to form them."[15]

It follows that the evolution of Descartes's formula from "cogito ergo sum" to "I narrate, therefore I am" cannot be reduced to the subject's construction of *content*. The second meaning, which is ultimately central here, has to do with the practice of linguistic and analytical competencies. On the syntagmatic level, narration requires the storyteller to work through contingencies and to create plausibility, synthesis, and conclusion. On the paradigmatic level, it requires formal, aesthetic considerations in the choice of words. The questions "Why?" or "What for?" and the ordering of parts into a before and an after strip events of their episodic nature, which saw them as rising up out of the past without explanation and disappearing into a future without consequences. The questions and the ordering of the parts compose time and give the subject, as both narrator and reader, orientation. They are the source of the plausibility that must be narratively achieved through the use of causal connectors and intentional markers—a plausibility that on occasion can also synthesize apparently heterogeneous elements. Awareness of the storyteller's unique perspective is also sharpened formally, as she wrestles with convention in her struggle to find the right linguistic expression. More simply stated, you come to understand yourself in the process of grasping what occurred and why.

The surplus value of "I narrate, therefore I am," in comparison to recently popular self-representation formulas like "I post, therefore I am" or "I share, therefore I am," lies in the cognitive activity that is involved. On Facebook and other social media, cognitive competencies, including analysis, synthesis, and formulation, are required only in specific situations and in a rudimentary way. They remain entirely unused when the report is automatic: "Roberto has shared a link . . . Michael, Antje, Eric, and 20 others like this." From the perspective of narrative psychology, such automatism leads to an utterly absurd variation on Descartes's formula: "It posts, therefore I am." The model of an insular and partly mechanical self-representation leaves behind a void that—this is the point here—is no less problematic than the economic and political exploitation of the accumulated data. The individual appears, above all, as the object of her history—and scarcely at all as the subject of its narration.

This constellation reminds us of the diagnosis of the late-modern individual's relationship to the self, according to which our era creates "an intensified compulsion to focus on the self and an urgently felt need for articulation, accompanied by a poverty of expression," and favors the "'de-temporalization' of life in favor of situational practices in regard to time and the self."[16] The preference for episodic over narrative self-perception that is suggested here will concern us in what follows. First, though, we need to ask to what extent social networks, as a means of identity management under conditions of accelerated relations to the world and the self, are the technological expression of this very contradiction between the compulsion to focus on the self and the failure of articulation. To what extent do social media make it possible, in a tradition-starved, radically flexible world, to decouple self-reflection, with its attendant gains in meaning, from self-representation? Or, to put it a different way: Does Facebook, as a technology for the permanent archiving of situations, cause the present to disappear by demanding and permitting no distance from it? The related counterquestion is: Does a narrative self-understanding that sees events (only) from the perspective of a given "system" really create sufficient distance from the present to be close to it in Agamben's sense of "true shared contemporaneousness," or does it, rather, fail to perceive the present in its actuality precisely because of its pregiven narrative perspective?

Before trying to answer this question, it is necessary to interrogate the claim that there is a discrepancy between self-representation and self-reflection. Wasn't it Facebook that first made self-description a central factor in our lives, even beyond the 5 or 10 percent of individuals who kept a diary anyway, especially when they were young? Don't millions of people experience themselves more intensely since Facebook appeared, starting with the choice of profile picture and welcome statement and responses to the list of questions and repeated with every status update and comment on the updates of other people? Doesn't the chronological sequence of events on a person's Facebook page already produce a certain narrative unity that possibly also sharpens her awareness of the way it all coheres? Doesn't Facebook itself, with its automatic "Say Thanks" collages and "Year in Review," with its monthly sequence of the most popular images and associated texts, teach our pictures themselves to tell a story? Shouldn't we be talking about a "narrative turn" instead of about the end of narrative?

The concept "narrative turn" points to the rediscovered power of stories before Facebook and irrespective of Facebook's role. It has been used in the business world for company identity and product marketing; in politics, as an effective means of earning voters' loyalty; in sociology, as a methodological defense against empirical social research; in medicine, as an alternative/traditional method of diagnosis; and in television, which hadn't dared allow itself this much focus on narrative for a long time. Even in literature, stories are once again being told.[17] The question that remains is how narratives in politics, economics, medicine, and research relate to the process of emerging (self-)consciousness in the sense of narrative psychology. Is there, along with the narratives of the specialists and the narrative-identity creation of businesses, also a return of narration in the concrete behavior of individuals toward one another and themselves?

All of this suggests—indeed, this is what is being argued here—that the popularity and ease of information exchange on digital media is displacing traditional forms of communication that have a strongly narrative character: the reflective diary, the letter reporting on events, the story that unfolds while looking at a photo album or watching an evening slide show. The thesis is that both the social and the technical *dispositif* of social and mobile media have a deleterious effect on storytelling. The smartphone's keyboard does not invite lengthy writing, and the logic of attention economy does not permit long reports on the social network.

In other areas beside these social media, we also find phenomena that work against the trend to rely on narratives. Some of these are once again a consequence of the new media. In participative journalism (tweets, smartphone videos, life blogs), for example, when the "truth of ordinary witnessing" replaces the voice of the reporter, it is undoubtedly a move away from the narrative style of the New Journalism. News is then no longer what actually happened (and, above all, why); it is how participants experience it. Narrative work is replaced by "truth beyond doubt," for no matter how inappropriately witnesses may report, the moments are always authentic—after all, the eyewitnesses are not actually reporters; they are the report! Ontologically, what is occurring here in regard to media is fundamentally a reorientation from the subjective mode of textuality to the objective mode of photography. At the same time, these changes can also be understood as a shift from the narrative principle to the database principle: The narration of a reflective reporter is replaced by disconnected mininarratives—a timeline of events unaccompanied by interpretation.[18]

The newsfeed principle of "always-on" journalism brings us back to Facebook, where the livestream also replaces comprehensive narratives. The shunting aside of narrative journalism relativizes talk of a "narrative turn." Further research is needed to determine whether narratives, today, play a greater or lesser role in individual lives.[19] We need to ask whether the explosion of communication via mobile media and social networks fosters or discourages the narrative turn in everyday situations. We need to ask whether the social and technological *dispositif* of these technologies and forms of communication fosters a culture of conversation in which listening and responsive questioning occur. Facebook, which for well-known reasons prefers information that is suitable for the database, must also be interrogated. We need to explore how the various forms of self-representation, from self-reports on questionnaires to status updates and automatic documentation of activities, are related to the process of narration. Do they encourage the communicative form of narration, or are they, as is suspected here, the expression of an *antinarrative turn* dressed up as narrativity?

*   *   *

Anyone can see that the protocols for self-description, with their forms to be filled out, do not require or encourage narrative competencies. They merely ask you to respond to the questions on the form (education, employment, home address, family members, favorite quotes, and "some details about yourself"). The self-description on the form is subordinated to the authority of the form, with its assumptions about what constitutes identity. In the case of Facebook, this includes statements about a person's favorite books, films, music, and athletes but (for example) not her favorite number, color, animal, mineral, season, or time of day. Every occasion for self-description that goes beyond the empty pages of a diary contains culturally determined implications and aims. To the extent that an identical questionnaire is employed for significant numbers of people, it aims at standardization. Autobiographical self-discovery, then, is thwarted in principle by the collective compulsion of the frame in which it takes shape.[20]

Compared to the multiple-choice segments, the fill-in segments "About You," "Favorite Quotes," "Religious Views," "Political Views," and "Life Events" naturally afford a certain freedom of self-description. Here, too, however, the parameters reveal how the employees of Facebook understand identity. For example, the heading "Life Events" contains five sections: "Work & Education," "Family & Relationships," "Home & Living," "Health & Wellness," and "Travel & Experiences." Each rubric contains further subdivisions. For "Family & Relationships," they are "First Meeting," "New Relationship," "Engagement," "Marriage," "Anniversary," "New Family Member," "New Pet," "End of Relationship"—and more recently also "Create Your Own." These segments almost read like the stations of a very normal life story, with the profound (and undoubtedly unintended) irony that neither the birth of a child nor the acquisition of a new pet could prevent the breakup of a relationship. In the segments themselves, there are blanks to fill in for "Who," "When," "Where," and "With Whom" and for more specific information (such as the name, type, breed, and gender of your pet). There is also the possibility to upload photos, and—"optional"—to tell a story about the event. The life events are—rather secretly—listed chronologically in the "About" section but are not connected or connectable in any other way. It is evident that even the section on life events, despite its partially narrative option, represents, above all, yet another form of data request, to be input in a database-friendly form.

A less formalized possibility for self-description is offered by the status updates and comments, which can be understood as "small stories" and from which, in some cases, the plotline of a bigger story could be pieced together, especially by the "friends" of the Facebook user who are acquainted with her offline.[21] However, to the extent that the updates usually take the form of spontaneous snapshots and are limited to unannotated information about places or activities, they do not cohere with Ricœur's concept of narrativity, and neither can they be understood as a kind of pointillist self-portrait, as has been claimed. For the problem here is not the narrative style (or lack of it) but the uncertain authorship of this kind of portrait. The sum of the collected "small stories" does not result in a pointillist self-portrait, which, despite the particular style of brushstrokes it employs, has been created intentionally with an eye to the result. Instead, they are more like what Zygmunt Bauman describes as "moments into which the pointillist time of liquid modernity is sliced." These moments are less narrated than noted during the moments of their passing, partly without the user even remarking on them, and are instead automatically registered by the technical frame.[22]

The "pointillism" of "narration" on Facebook is not just an extension of the episodic perception that Bauman already remarked on when it comes to postmodern subjects. It is simultaneously an inevitable consequence of the technical frame. The interface constrains coherent storytelling by not foreseeing or permitting an internal link between the events on a person's own Facebook page. This omission is quite astonishing. After all, links are part of the fundamental technological structure of the internet, and "connecting" is integral to its philosophical self-concept. On the other hand, it is not really so surprising that Facebook does not encourage any narrative activities on its front end, if the central goal of all this user activity is the data analysis being carried out at the back end. Thus, instead of causal relationships created by autobiographical narrators, we find only chronological connections based on the temporal sequence of the events. The technical *dispositif* creates a situation in which the individual subject/object of the updates no longer creates the narrative order of her life while writing but more or less unconsciously produces it while living it. The work of narration is thus simultaneously actionistic and postactive: By letting actions speak for themselves at the very moment when they occur, the narration no longer occurs at the level of presenting data but instead at the level of its production. This reduction in narrative consciousness was initially advanced by Facebook's decision, in December 2007, to abandon its original status-update prompt "*Username* is . . ." This meant that users no longer had to assume the rather distanced and reflective posture required by the—quite unusual (almost avant-garde)—practice of speaking about themselves in the third person.

From the perspective of narrative psychology, such undermining of narrative consciousness appears to be a loss, but according to the logic of the participation culture that characterizes the Web 2.0, it is just another example of democratized communication processes. The shift in narration from the autobiographical subject—including the self-censorship that goes along with it—to the more or less automated collection of status updates and reports and the contributions by the networked public means the end of rule by experts, even when it comes to autobiography. What remains to be ascertained is the likelihood that this narrative activity will occur on the part of the public. While we cannot exclude the possibility that Facebook users create an overarching narrative image based on the various data available on a Facebook site (and offline), we do need to ask to what extent the habitual practice of isolated statements obstructs the creation of narrative structures during the reception process and whether, in principle, it encourages an antinarrative view of the world.

The question of the psychological implications of the technical *dispositif* should also be posed with reference to other forms of everyday storytelling: letters, telephone conversations, or shared conversations in physical proximity. If self-representation, which in face-to-face personal contact is dialogical, reflective, and narrative, is increasingly shifting to the realm of digital communication, it is also increasingly subject to the rules of attention economy and self-management that apply there. The next question—whether status updates on the social network make personal stories in the offline realm superfluous or, on the contrary, serve as occasions for responsive inquiry—requires empirical investigations that are outside the scope of this study. But the topic of delegated narration should be further explored.

DATABASES AND MECHANICAL NARRATORS

Facebook outsources narration to readers only potentially and in a limited way. The actual narrator does not sit at the front end of the interface, where users read, write, post, and imagine, but at the back end, where algorithms analyze the data that has been collected. At the back end, Facebook, as is sufficiently well known, is a giant database that collects data sets for every user under some sixty categories.[23] There, the data points that have been disaggregated by the questions on the forms are reassembled twice: into a profile of the person in question and into networks of relationships among the many. The best basis for this work is not the subjective construction of a person's own history but its "raw material," as the objects of photography were dismissively termed in the nineteenth century. In the jargon of Big Data, this is celebrated as "raw data." Any narrative treatment of the data at the front end of the interface represents an obstacle for this project, since, from an information-theoretical perspective, narrative attempts to extract the "real" truth of the whole from the "literal" truth of details can only result in distortion. A complete lack of narrative connection is therefore the best guarantee that a given data set is all-embracing: If there is no story to be told, there also cannot be any data that fall outside it.

The busywork of algorithmic narration reaches all the way to the front end, where user data is accumulated according to specific criteria. The "activity log" includes the user's likes and comments, together with other people's photos on which the user is tagged, while the function "Suggested Friends" collects messages to or from a specific Facebook friend. There is also the "Say Thanks" service and the "Year in Review." In assembling all this data, so far, the system does not exhibit very much narrative energy; it merely presents the explicitly declared (and marked) links between two individuals, not the "deeper" relationship of two Facebook friends who, for example, have watched the same video and afterward read the same article. That the system is not capable of producing this kind of correlation should hardly be assumed. Instead, we should ask when applications are going to make the more complex forms of data gathering that are taking place at the back end available to users on the front end.[24]

Facebook's algorithmic storytellers are symptomatic of the development toward postactive narration in Facebook society.

Other technologies produce images that involve no personal involvement and reports that have no human authors. Narrative Clip, for example, is a small portable camera that can be worn on the lapel and produces a photograph every thirty seconds, thus relieving the subject of the need to make any decisions during the taking of autobiographical photos—all she needs to do is make a selection of usable photos from among the day's yield. The process encourages the exposure of the optical unconscious that Walter Benjamin identified as a characteristic of photography. Another program of this kind is Narrative Science, which embeds concrete data within a narrative scaffolding and produces reports (for example, in the realm of sports or finance) scarcely distinguishable from those prepared by human authors. The question that concerns us here is, naturally, whether at some point a program of this kind could also be adapted for Facebook, employing existing tags and visual references (which can be collected by facial-recognition software) to tell a story that informs the relevant subject what she has actually experienced and who she really is. Could the technology make the objectivity of the story, after having extracted it from the control of the subject, valid and persuasive for that person? Or is the "narrative turn" on Facebook, Narrative Turn, Narrative Science, and other sites offering automatic narratives a turn toward narration that no longer has anything to do with its hero, because its actual goal is the investigation of proclivities, interests, and activity models?

Are people, at least, still narrating their own history with the help of the diary apps now being offered in increasing variety by the market? These apps are not merely technical enhancements to journals that allow them to add images, sound, and options for copying and sharing. Nor are they merely the next step after Word, which, while it functions multimedially and easily allows copies, basically operates on the model of the blank page. Diary apps are the negation of the diary because they bring about the event of self-description in a way that transcends narration. The limited space for text, the uninviting writing tools, and the user context already speak against the contemplative pause and personal report that traditional diaries once offered. For example, the smartphone's small screen and narrow keyboard are hardly conducive to lengthy writing. Besides, in the apps' interface the reporting function generally plays a lesser role compared to photo uploading and factual reports. Moreover, the installation of this kind of app on a smartphone favors a kind of diary "on the go," which corresponds to the acceleration of modern life, rather than countering it with a moment of stillness.

Naturally, the app marketers don't find anything amiss in a "diary" that requires no leisure, instead praising the relatively effort- and thoughtless report on the Now as an advantage of their product: "With Momento in your pocket you can write your diary 'on the go'" (momentoapp.com); "beautifully automated. Effortlessly remember every single day of your life" (roveapp.com); "makes remembering effortless, beautiful & fun" (hey.co). It is ironic that precisely the app that at first blush seems most invested in speed is the one most likely to encourage retrospection: "1 Second Everyday" asks users to add a single image or brief video each day to a history of the month or the year; thus it at least requires some reflection as to which image best represents the day. The focus of diary apps, as with social networks, is on database-friendly information (time, place, participants, tags) linked to photos and a brief text. Instead of the coherent narration of experience, we get an episodic report; instead of a reflective diary, a logbook for short answers to pre-programmed questions that can be responded to in the midst of other activities: How did you sleep? What are you doing? How do you evaluate your current creativity, on a scale of 1 to 100?

The database logic behind all this, advertised with a view to the possibility of keyword searches, demands activity that is less narrative than identifying. Users are meant to mark places, people, and events: "Food," "Dreams," "Business," "Friends," "Vocation," "Love," "Joy," "Idea," and "Movie" are all set to go on Diaro, with the option of adding your own criteria. The idea is to organize activities into pregiven categories: Optimized offers "Creativity," "Routine," "Pleasure," and "Health," without the option of associating a single event with more than one category. Like Facebook, some of these apps (for example, Momento) also automatically integrate external communications (on Twitter, YouTube, Facebook) and reactions to them, repeating the action-istic, postactive method of storytelling described here. Like Facebook, these apps permit no links between entries. Here, too, technical updating of the old self-observation technology of the diary takes place in the interest of the database paradigm, with-out concomitant updates in the interest of the narrative para-digm. Facts supersede links—a technical decision with explosive social impacts.

The answer to the question previously posed is: On diary apps, an individual is still the narrator of her own story only in signifi-cant entanglement with the logic of the database. The goal of these apps is not to work reflectively through lived events as part of hard-won life experience; it is instead a maximally clear, fac-tual report, a kind of spontaneous eyewitnessing of oneself. Thus, the app Reporter markets itself with the slogan "Snapshot your Life," promising that by filling out brief daily question-naires (What are you doing, when, with whom, for how long?), its users will shed light on the nonmeasurable aspects of their lives. Its name already harks back to the phenotype of "cool" photographic observation as it has been described, promising a database-centered self-description shorn of any narrative self-deception. It is a self-description that, after the information "This is your relationship to . . ." in the "Say Thanks" collage and "This is what your year looked like" in the "Year in Review," proceeds purposefully toward the result: This is who you are.

The next stage in nondescriptive witnessing is already fore-seeable in new software and hardware that is guaranteed to record and share every single thing a person sees and experiences: Twitter's Periscope, Google's Glass (or whatever its future equivalent will be), and Facebook's Oculus Rift. Zuckerberg is convinced that virtual reality (VR) technology will bring the next great turn in the medial ecosystem, comparable to the effect of the smartphone on desktop computing. This time, he wants Facebook to play a defining role in the change. At F8, Facebook's developer conference, in March 2015, the integration of immersive 360-degree videos via the Oculus Rift ("spherical video") was one of the main projects for the future. This recalls scenarios of the unbroken, automatic recording, replay, and sharing of experience as envisioned in late 2011 by the British science fiction TV series *Black Mirror*, whose episode "The Entire History of You" depicted technology that let people record their inner experiences audiovisually, like an external camera that also grasped things the individual wasn't consciously aware of. Users could play these experiences back on an external screen and share them with others. Zuckerberg's plans in this regard have been clear at least since his Q&A session on July 1, 2015:

> We'll have AR and other devices that we can wear almost all the time to improve our experience and communication. One day, I believe we'll be able to send full, rich thoughts to each other directly using technology. You'll just be able to think of something and your friends will immediately be able to experience it too if you'd like. This would be the ultimate communication technology.[25]

If all goes according to Zuckerberg's plans, the future of self-presentation will transcend any conscious (linguistic) representation. Whatever is to be communicated communicates itself without the distortion of passing through the reporter. This is why Facebook is simultaneously working on artificial intelligence systems that recognize all the elements in an image. In this way objects—particularly if the automated image is generated quasi-unconsciously—can present themselves while bypassing the subject. It is certainly not surprising when this loss of self-observation—as observation *by* the self—is presented by Zuckerberg as a gain in experience and communication. The automatization of the report and the expulsion of the subject from self-narration are the logical consequences of the transparency doctrine: Human beings, consciously and unconsciously, always want to conceal something; only machines have an objective interest in knowledge.[26]

The real-life precursor of the future Oculus automatism is the increasingly popular app Snapchat. However reliable the promise may be that the images sent via this app disappear a few seconds after they are viewed, it has already contributed significantly to a shift of communication toward the nonverbal realm of image sharing. For many users of Snapchat, the guarantee that the snapshots will self-delete is less important than the possibility of engaging in visual communication that is as spontaneous and banal as possible. Instead of saying what you are doing and how you feel, you send a snapshot. The descriptive-communication model of language gives way to the indicative model of images; the individual perspective on things is reduced to an effect of the camera. With corresponding image-recognition software, it is possible to analyze billions of situations while bypassing their actors and reporters. As long as the images are not also erased from the back end of the interface (which is what we should assume), this method of communication represents a goldmine for data collectors, which may explain why the two founders of Snapchat did not sell their app to Facebook in 2014, although they were offered three billion dollars for it, but went public in 2017 for $29 billion.

The irony in all of this is that an application that started out as a means of protecting the private sphere is helping impose the transparency doctrine, in the sense of the self-reporting of objects and events. It is equally ironic that although the snapshots of the most recent twenty-four hours are collected under the title "My Story," this "history" is less available to the person who created them than it is to the system. By evening, committed Snapchatters scarcely know any more what they have photographed and shared during the day. What they do recall is merely the pattern that guides their actions in each case: The "pre-gym selfie" is followed by "ready for the gym" and "in the gym" pictures and then by "after running 5 miles" and "post-gym food." To the extent that Snapchatting replaces texting—not to mention diary writing—it means not just the disappearance of the images but also the evaporation of the person's own history. Now only the algorithms at the back end know better—and anyone who has access to them.[27]

As silly and banal as the results of the "Thank You" collages and Narrative Clip may be at present, however innocent an app like Snapchat may seem at the moment, we are experiencing the beginning of something that in a few years may fundamentally change our narrative self-understanding. The tendency is toward an automatic immediacy of reporting, which is actionistic and simultaneously postactive. As with "news as it happens," events are recorded the moment they occur, and, as with instant journalism, the lack of distance of this type of autobiographical "writing" bars the path to a reflective view of things. Today, there are undoubtedly more people filling in a Facebook page or a diary app with data about their life than there once were people writing in a diary. But the quality of self-observation on social networks is significantly different from that of diary writing. Subsequent reflection is replaced by spontaneous reports, mostly oriented to externally imposed questions and criteria. The famous self-discovery of the diary writer moves away from reflective exploration and turns into the question of how the individual fits into a prescribed pattern of recommendations and expectations—assuming the report is not automatic anyway and hence completely removed from consciousness.

The contradiction and active opposition between the database and narrative models of information processing were already subjects of discussion fifteen years ago. At that time, the database version was seen as the form of self- and world perception that was most adequate for our era.[28] That the structure and orientation of the database are equally determined by cultural assumptions can already be illustrated, in an odd way, by Facebook's questions about life experiences, where the rubric "weight loss" is not accompanied by a parallel rubric "weight gain," and the question is asked "with whom" (one lost weight) but not "for whom" or "against whom." Self-tracking apps show similar oddities. It is evident that categorizations have a decisive influence on how reality is perceived and are therefore neither epistemologically nor politically innocent. "Categorization is a powerful semantic and political intervention: what the categories are, what belongs in a category, and who decides how to implement these categories in practice, are all powerful assertions about how things are and are supposed to be."[29] But the "narrative pollution" of the database changes nothing about the fact that in the twenty-first century this explanatory model is becoming ever more central to our perception of ourselves and our world. Facebook is an example of this tendency not only because it prefers to tabulate the results of questionnaires and blocks narrative forms of presentation at the front end of the interface. The algorithm at the back end that decides whether status updates from our friends appear in our newsfeed also prefers photos, videos, and links, rather than textual entries, which are less compatible with databases.

The key question is what problem the move to databases is designed to answer. The Russian-American media theoretician Lev Manovich throws three names and slogans into the ring: the death of God (Nietzsche), the end of grand narratives (Lyotard), and the arrival of the World Wide Web (Tim Berners-Lee). The world, he argues, appears as an endless unstructured collection of images, texts, and information; hence it is only logical for us to model it as a database.[30] Manovich actually fails to provide the underlying reasoning that would support his thesis, but his list clearly indicates that the database is to be seen not merely as a consequence of technological inventions (the web) but also as the result of cultural development (end of grand narratives). We may certainly see the database as the overthrow of postmodernism's relativist model of knowledge or as the victory of quantitative certainties over narrative, theoretical, or ideological constructions. More paradoxically, in a formulation that harks back to the origin of the concepts: The database is the return of narrative as number.

This supposition is supported by the recognizable effort to break information down into calculable units, something that, as the example of nanopublications illustrates, can take verbal as well as numerical form.[31] The shift from narrative to numbers is even more evident in forms of contemporary self-knowledge such as the Quantified Self movement, whose slogan "Self-Knowledge Through Numbers" shows an unmistakable mistrust of narrative self-observation. Self-tracking, or "scanning," has also been viewed as an extension of confession or psychoanalysis and as another form of "egotistical cultural practices" that humans invent in order to discover their "true self."[32] This, however, would be to deny the difference between a process of self-observation through scanning and one that is reflective, or, more pointedly, to confound the self-exploration of an athlete with that of Augustine. It would be equally problematic to see the Quantified Self movement as an expression of the "care of the self" that Foucault, in the early 1980s, discussed as the ancient Greeks' art of living and recommended to his contemporaries under the rubric of an "aesthetics of existence." The objective of this care and this aesthetics was a self-conscious and self-determined life under the sign of a person's own needs and values. The care was directed equally toward body and soul, whereby exercising the body also always serves to care for the soul: "as physiotherapy that in truth is psychotherapy."[33]

The primarily *physical* self-optimization of the Quantified Self movement turns Western culture's hostility toward the body into hostility toward the mind and spirit when it allows the commandment "Know thyself" (once an inscription on the Temple of Delphi and later the main ambition of self-exploratory seminars and trips in the 1970s) to degenerate into an obsession with self-measurement. Corresponding apps and social networks provide the necessary technologies for collective control of jogging and pulse rates or of movement, sleep, and eating behaviors. One of the movement's heroes is Nicholas Felton, whose "Annual Reports," appearing on the internet since 2005, present important data from his life with statistical exactitude and a pleasing design that indicates how often he took a subway, taxi, bus, airplane, ferry, or ski lift; how often he visited a museum or the gym; and how many books he read, with how many pages. Felton is an extreme symbol of the shift from reflective narrative (about the themes of books or the experiences had on the trips) to detailed numbers (calculations of pages read and miles driven or flown), for which Felton employs the concept "numerical narratives."[34]

The self-knowledge favored here is based on numerical values and correlations that, while they must ultimately also be interpreted, are nevertheless—at least this is what the self-trackers assume—more reliable than self-description. "It is possible that the data-mapped, virtual self offers a more accurate picture of who we really are than the subjective stories we tell," one observer holds, suggesting we should befriend the "encountered" alter-ego: "We can learn to love the data-mapped self that reveals our real behaviours, in all their complex, contradictory, hypocritical glory." If this encounter is described not as alienation but as the consolidation, using physical means, of a psychically destabilized ego—"In self-tracking, we are literally trying to keep track of the body, to rephysicalize it, in an adaptive reaction to the ungrounding of the self in contemporary life"— then the assumption that objective data is reliable is also simultaneously revealed as a method for managing anxiety during times of rapid changes and increasing uncertainty.[35] The obsession with data becomes an ersatz action that the individual expects will provide a new source of orientation. The identity crisis that results from the loss of narrative forms and formats (and, above all, from the lack of trust in them) is managed by a methodical change in the epistemological model from words to numbers. Against the reintegration of the individual in narrative wholes, the self-trackers bet on self-assertion through self-quantification, with the idea they can thus avoid being implicated in any grand narratives other than that of the number itself.[36]

Naturally, the methodical point of departure for numerical self-knowledge is also culturally determined. The practice of self-measurement is coherent with the general social trend, as an adequate control mechanism in the era of digital media. Therefore, the self-quantifiers' so-called body hack is not an act of rebellion against a governmental or economic system but an attack on their own body in order to create more data about it and put that data at the disposal of scientific but ultimately also governmental and economic interests. To avoid conspiracy theory foreshortenings, both of the motivating factors behind this control-friendly hacking should be recognized. On the one hand, there is naturally an interest "from above" in the biometric datafication of the subject, not least because this data can be applied to the post-Fordist labor process, under the euphemism of gamification, as an imperative both for the collective struggle to get ahead and for individual self-optimization. On the other hand, however, there is also an interest "from below" in technologies of self-tracking that make it possible to realize traditional values, such as healthy nourishment and physical exercise, with the help of external methods of measurement and motivational assistance. Thus, the datafication of the subject, like her "wiring" into the "internet of things," is a phenomenon that belongs to the cultural logic of modernity. It is not generally enforced against the individual's will and is also not always contrary to her interests.

The turn to numbers is, admittedly, only perfect once the instruments of measurement are directly connected to the body of the subject, bypassing her awareness. Only the automatic collection of data protects it from being manipulated. The examples that have been cited are "frictionless sharing" on Facebook and the objective snapshots of Narrative Clip. Additional forms of machine-generated self-representation include the Foursquare app Swarm, introduced in 2014, which offers a "Neighborhood Sharing" function to automate the "Check-in" that communicates a person's location to the social network. Then there is the app SpreadSheets, which automatically records data on sexual activity, using an accelerometer and a microphone (its predecessor Bedpost required entering this information by hand).[37] The two apps exemplify the trend to document individual behavior based on data provided by the user's *body*. Cultural content, as expressed in words, is negated by the nature of the body, which offers information about itself via technologies of digital measurement that bypass consciousness. Thus we produce traces that are unavailable to us personally, even if we did, at one point, turn the mechanism on and were conscious of our actions in each concrete situation. But in the end, it is the algorithm that can provide information about what places we visited three years ago, what we did a year ago on this or that day, or which status updates we "liked" a month ago. It is the algorithm that knows what we are up to, to the extent that knowing means knowing about the external data that our life yields.

POSTHUMAN SELF-DESCRIPTION

Jean-François Lyotard problematized grand narratives long before the internet and databases had become symptoms of cultural change. Accordingly, Lyotard's reception, at first, did not include references to the new media, although it did focus on the work's philosophical and narratological implications. Thus, the German-Korean philosopher Byung-Chul Han characterizes Lyotard's accentuation of the *that* (something occurs) rather than the *what* (occurs) as a "turn towards *being*": "In the age characterized by narration and history, being retreats into the background in favour of meaning. But when meaning retreats in the course of de-narrativization, being announces itself." The "dissolution of the narrative chain" frees perception from the "*chains* of narration", that is, narrative coercion, to events in the proper sense of the term. The event—contradicting Gatterer's bon mot of 1767—remains the event even when it is not part of any (narrative) system.[38]

In the context of the narrative theory of meaning, which Han represents, the moment that has been freed from the fetters of narrative possesses "profoundness of being" but no "profound meaning": "its profoundness only concerns the pure presence of the *There*. The moment does not re-present . . . The *There* is all it contains." This pure presence, this phatic communication with no aim other than itself, is precisely the problem for Han. Against Lyotard's opinion that the "end of narrative time" makes it possible to come close to the "mystery of being," Han argues for the "nihilistic dimension" of such a perspective: "The decay of the temporal continuum renders existence radically fragile. The soul is constantly exposed to the danger of death and the terror of nothingness, because the event which wrests it from death lacks all duration."[39] Modern humans' experience of time, Han continues, is a "rugged, discontinuous event-time": fullness without direction. We are constantly starting over, channel-hopping through "life possibilities," precisely because we are no longer able to carry a possibility through to the end: "The time of a life is no longer structured by sections, completions, thresholds and transitions. Instead, there is a rush from one present to the next."

The finding can be expanded phylogenetically, as the problem not only of the individual but of humanity as a whole: "The end of history atomizes time into point-time . . . history gives way to information. The latter does not possess any narrative width or breadth." Precisely because information is a phenomenon of "atomized time" or "point-time," it must strive all the more hysterically to fill the voids between these points, which can no longer be experienced as part of a narrative line. Thus, perception is always supplied with new or drastic materials. Atomized time permits no contemplative lingering. Not knowing where we are going leads to a narrative stasis, which is camouflaged with a flood of events and which, with Heidegger, could be disqualified as an absence of dwelling (*Aufenthaltslosigkeit*) in a meaning-deprived, sped-up sequence of mere happenings: We are channel-surfing ourselves through the world. Since the death of God, humans naturally no longer redeem themselves from the lack of dwelling in earthly existence by means of cosmological experience, for example by living toward the fulfillment of divine time. Yet at the same time, since the "end of history" has also rendered world-changing heroism impotent, humans are no more heroes of time than the monks in the *Annals of St. Gall* once were. Time, fragmented into its individual moments, is no more, now, than what must be *experienced* or, depending on one's perspective, *endured*.[40]

For Han, distracted busyness is no more a solution to the crisis of existence than was Lyotard's existential profundity. Han turns back to Heidegger's construction of contemplative lingering and votes for a "return-to-self": "The end of narrative, the end of history, does not need to bring about a temporal emptiness. Rather, it opens up the possibility of a life-time that can do without theology and teleology, but which possesses a scent of its own. But this presupposes a revitalization of the *vita contemplativa*."[41]

Han's construction repeats a figure of thought that is found not only in Heidegger but also in Kracauer, who, at the time of the publication of Heidegger's *Being and Time*, was writing about "metaphysical suffering from the lack of a higher meaning in the world, a suffering due to an existence in empty space."[42] For Kracauer, there were three kinds of alternatives to the "cult of distraction" that emerged as a reaction to metaphysical homelessness: principled skeptics (also called "intellectual desperados"), "short-circuit people" (who fled "headlong" into a new belief), and "those who wait" (and who tried to achieve a "relation to the absolute" by means of a "hesitant openness, albeit of a sort that is difficult to explain").

Kracauer did not specify what "those who wait" were aiming at with their "hesitant openness." But elsewhere he made clear that their stance was preceded by a refusal of distraction and a willingness to be bored: "The world makes sure that one does not find oneself. And even if one perhaps isn't interested in it, the world itself is much too interesting for one to find the peace and quiet necessary to be as thoroughly bored with the world as it ultimately deserves."[43] Kracauer's early critique of the quality of being "interesting," which, long before Facebook, was evidently presenting and imposing itself as something that it really was not, is worth noting. His suggestion of a way to fight back prefigures Picard's praise of quiet and implies abstinence from the media: "But what if one refuses to allow oneself to be chased away? Then boredom becomes the only proper occupation, since it provides a kind of guarantee that one is, so to speak, still in control of one's own existence." Kracauer's mandate refers to the passage from Nietzsche's *Thus Spake Zarathustra*, with which Han also concludes the German edition of his book, in which Zarathustra criticizes all those "to whom rough labour is dear, and the rapid, new, and strange": "If ye believed more in life, then would ye devote yourselves less to the momentary. But for waiting, ye have not enough of capacity in you— nor even for idling!"[44]

The world to which Han and Kracauer recommend waiting and laziness as the royal road to contemplation has become unfamiliar with phylogenetic stories that could have offered an ontogenetic foothold. Postmodern man no longer experiences himself as part of a social project. He is not a pilgrim on the "path of progress" toward himself and the deeper meaning of life; he is a tourist who doesn't want to be determined by the past or constrained by the future, a "flexible" man with a "situational identity" who "lives at the vanishing point of individualization and acceleration" and has forfeited the "claim to (diachronic) continuity and (synchronous) coherence." He lives under the "impression of *racing stasis*: things change, but they do not develop."[45] This is the more recent summary of the postmodern subject's loss of narrative that Bauman similarly observed twenty years earlier: "The overall result is the fragmentation of time into episodes, each one cut from its past and from its future, each one self-enclosed and self-contained."[46]

What is decried here as the loss of a narrative home for the self, other commentators celebrate as the lightness of a "thin subject"—a radically ephemeral self that not only exists separately from its lived experiences but essentially also vanishes along with the event that has just been experienced, and reappears with the next one. This perspective undermines the theory of psychological and ethical narrativity according to which humans only experience their life when they tell it to others and themselves, allowing them to develop a responsible personality. It also challenges the model of the *diachronic* self, which emerges experientially in the coming together of past, present, and future, by positing an *episodic* type, which always lives in and understands itself exclusively in relation to the present: "One has little or no sense that the self that one is was there in the (further) past and will be there in the future, although one is perfectly well aware that one has long-term continuity considered as a whole human being." This does not mean that the episodic type lives without any memory of the past. But the memory occurs without any narrative passion. Yesterday is unreflectively present in today the way the past rehearsals of a musician are present in an actual performance.[47]

The critique of the identity concept of "ethical-historical-characterological developmental unity," which in a sense transfers the modern concept of development to the individual, puts paid to Heidegger's assertion that episodic personalities are necessarily "inauthentic" in their experience of temporal existence: "But I think that the Episodic life is one normal, non-pathological form of life for human beings." The proposal that is developed in response—of a life lived in the moment—is troubling to all those who, with the help of traditional criteria like identity, authenticity, or coherence, seek to describe a society in which other values (hybridity, change, momentariness) have long since come to determine the actions and self-understanding of individuals. It opens a positive perspective on the "presentism" of the (post)modern subject, on life in the now in the context of mobile media and social networks, and it resists the model of psychoanalysis (which provides the foundation for the ethical-narration thesis), according to which moral growth happens through reflection that simultaneously entails the overcoming of the narcissistic id by the social ego. Thus, not least of all, the construction of an episodic identity relativizes the critique of digital media's antinarrative *dispositif.*[48]

In a 2014 essay, the French grandmaster of autobiographical research Philippe Lejeune bemoans the decline of autobiographical identity in the era of acceleration and predicts coming forms of autobiographical writing on social media that will be incoherent, hypertextual, and multimedial.[49] Autobiography on Facebook is, in fact, incoherent, hypertextual, and multimedial. It is simultaneously posthuman, on all three levels of possible authorship: users, network, and algorithms. The sovereignty of the autobiographer is already fundamentally compromised when, following the "authority of the form," users adhere to the value assumptions and standardizations on lists of questions and categories. It is further weakened by the montage that is created when a person's own status updates are combined with the comments and status updates of friends. It is utterly lost when the algorithm becomes a "ghostwriter" with plans of its own.[50]

Naturally, Facebook is not the first to challenge the sovereignty of the autobiographer. The discourse of postmodernism already introduced external entities as the actual actants and affirmed the "death," or disappearance, of the author, since the ego is not the sovereign source of its feelings and thoughts but merely the point of intersection of its discourses. In both postmodernism and posthumanism, the subject lacks agency and self-determination. Yet we should not overlook the different natures of the heteronomy at work here. In the postmodern context, the subject's competitors for sovereignty are human actors (relatives, acquaintances, shapers of discourse present and the past), while in the posthuman context the competitors of the subject are technological: algorithms. While postmodern autobiography is determined (or "distorted") by the *internalized* perspectives of the culture to which a person belongs, in the case of posthuman autobiography "alien authors" (the network and algorithms) take over the writing, which then occurs *outside* the consciousness of the subject. Unlike the posthuman subject, which transfers its role as an autonomous actor to software whose decisions it cannot control, the postmodern subject, despite its heteronomy, maintains authority over its actions and identifies with the perspectives that are presented to it to the extent that it adopts them. The difference is the internalization of the heteronomy in writing (postmodern), as opposed to the outsourcing of the writing itself to heteronomous sources (posthuman).

Rather than making a premature claim of continuity, we must therefore emphasize the new quality of the subject's disempowerment, which is now a disempowerment *of* rather than *by* culture. The dethronement of the autobiographical subject (as narrator) by the algorithmic narrator is simultaneously a "liberation" of the autobiographical subject (as narrated) from cultural heteronomy. For the algorithmic narrator operates independently of the predetermined assumptions of cultural value that are inevitably manifested in the reports of human narrators. It is true, as software studies emphasize, that codes and protocols are, in principle, culturally determined, but the example of the Foursquare app Swarm sheds light on the difference at stake here: While the human narrator will skip the "Check-in" at certain points because he finds it unimportant or discrediting, the automatic place identifier ascribes no values and allows no concealment—unless, of course, the user has programmed it to do so. The "algorithmic auto/biography" that comes into being in the posthuman mode of writing is the "blackboxing of the self."[51]

This automatism, it must be said, is liberation only if the report's precision and accuracy are valued more highly than the activity of making a report, in other words, if the performative act of narrating (with its practice of reflection, necessarily accompanied, as it is, by distortion) is factored out in favor of greater objectivity. The discussion of narrative psychology pointed to the problematic nature of this approach and emphasized the necessity of narration as a praxis of linguistic and analytic competencies. This praxis, too, should be further explored and problematized as a form of heteronomy of the subject—a process of disciplining enforced by means of cultural expectations that are to be fulfilled. The more negative the resulting judgment turns out to be, in this respect, the more readily some people will welcome posthuman narration. This response has been prepared, in turn, by the critique of narration as a form of world and self-representation.

Parallel to the death of the sovereign author, around the middle of the twentieth century narration began to be accused of betrayal. A coherent story, it was claimed, creates the illusion of order and reduces reality to pure logic. This insight, injected into the literature of French existentialism (Camus's *The Stranger*, Sartre's *Nausea*), was radicalized by the *noveau roman*, which fragmented reality and made it appear incoherent. Impelled by similar concerns, Marxist and poststructuralist writers criticized the conservative character of the genre of autobiography as the false appearance of an individually determined, coherent, and meaningful life. In this situation, when narrative incoherence, discrepancy, and confusion are promoted as being more true to life, the development of historiography, as it was described at the beginning of this chapter, makes another about-face. Where it had previously evolved from an accumulation of facts lacking perspective into a causal concatenation of events, historiography now turns back again, as narratives cease to be understood as "actual, inner" truth, the way Wilhelm von Humboldt had seen them, and now appear to be nothing but "euphoria" and the creation of "serenity" through order.[52]

In light of this critique of the illusion of coherence, posthuman practices of narration on social media can also be understood as a radicalization of postmodern poetics. The author—in postmodern thought still identifiable as an enunciating web of quotations from innumerable cultural sites—is further reduced until it becomes a merely *mixing* web of experienced events, composed of data administered by mechanical narrators. If the portrait that emerges in this fashion is experienced as an alien self, this, in turn, recalls postmodern concepts of identity that conceive the encounter with one's "own foreigner" as fostering the disintegration of the self in a way that encourages tolerance. The disempowerment of the self on Facebook—initially problematized as a loss of narrative engagement, then relativized by the rehabilitation of the episodic type of identity—ultimately appears as rescuing the self from narration's techniques of self-deception through the use of "unimpeachable" methods of data collection. This "human" aspect of the posthuman will continue to occupy us in the following chapter, as we turn to the negative aspects of narrativity for individual and collective identity formation.[53]

Let us, for the moment—less as a conclusion than as a prognosis—keep in mind the three-step evolution leading up to the posthuman narration of the self: (1) from words to numbers, when description is replaced by statistical information, as demonstrated by the example of the Quantified Self movement; (2) from mechanical to automatic processes, when the inputs are no longer consciously entered by the subject but are involuntarily provided by the body or, as in the case of Snapchat, emerge spontaneously and more or less unconsciously; (3) from option to duty, when the creation and analysis of data is no longer initiated by the person who produces it but is forcibly imposed or secretly undertaken by employers, insurance companies, or government agencies.

Let us also keep in mind the challenge that the episodic identity type, as confirmed by many contemporary observers, poses for the narrative self-understanding of previous generations. In Douglas Coupland's 1991 novel *Generation X*, one of the main characters says, "it isn't healthy to live life as a succession of isolated little cool moments. 'Either our lives become stories, or there's just no way to get through them.'"[54] What Generation X still cared deeply about—giving life, which at the time already seemed like an aimless collection of insignificant events, meaningful coherence—may have forfeited all relevance for Generation Y, the "Facebook generation." A dispassionate review must ask what the consequences are likely to be when the narrative wholeness of life disappears, history degenerates into mere information, and existence becomes nothing but the rush from one present to the next. The question is how the loss of individual and collective storytelling changes the way not just individual people but humanity as a whole deals with the past and the future and with others. It is a question about the political consequences of episodic identity.

3

DIGITAL NATION

It is not enough, then, to set idle chatter in opposition to the authenticity of the spoken word, understood as being replete with meaning. On the contrary, it is necessary to discern the conversation (and sustaining) of being-with as such within chatter.

—Jean-Luc Nancy, *Being Singular Plural*, 1996

Never before has an age been so informed about itself, if being informed means having an image of objects that resembles them in a photographic sense." What sounds like a commentary on our ubiquitous production and reception of images is actually Siegfried Kracauer's description of what he observed in 1927, when, as he noted, the world seemed to have taken on a "photographic face" and strove "to be completely reducible to the spatial continuum that yields to snapshots." Almost immediately, the doubt that comes through in his "if being informed" clause leads Kracauer to the opposite conclusion:

Never before has a period known so little about itself. In the hands of the ruling society, the invention of illustrated magazines is one of the most powerful means of organizing a strike against cognition. Even the colorful arrangement of the images contributes to the successful implementation of the strike. The *juxtaposition* of these images systematically excludes the contextual framework that discloses itself to consciousness.

Kracauer's doubt does not derive from the fact that images in illustrated magazines absorb the reader's attention in a glamorous fantasy world, as suggested by the photograph of the film diva that Kracauer invokes at the beginning of the essay. The source of the problem precedes that distortion. It is located in the medium itself, based on the fact that "in photography the spatial appearance of an object is its meaning," whereas, on the contrary, "in the artwork the meaning of the object takes on spatial appearance." Hence: "the two spatial appearances—the 'natural' one and that of the object permeated by cognition— are not identical."[1]

Along with the questions raised here about photography as a medium, a new one arises: Who decides the meaning of things? Is it humans or the things themselves? To perceive things at the level of their evidence, Kracauer argues, means to be prevented from gaining access to their truth. Baudrillard, later on, will frame this constellation even more sharply as the decision between the "philosophy of the subject" and the "anti-philosophy of the object."[2] Photography is a "means of organizing a strike against cognition" not only as a result of the intentional distortion of reality that happens in posed photos, where (viewed from the perspective of media theory rather than ideology critique) "the meaning of the object takes on spatial appearance," as in a painting. This is the kind of meaning the film industry, following its logic, would like to create for the diva, for example. But alongside the cognitive distortion that is carried out via the medium, there is a second cognitive betrayal that derives directly from the medium itself: the suppression of any subjective perspective by the camera's own vision, which reduces "truth" to the "naked" appearance of things in space. With the (apparent) self-presentation of the objects, the human observer finds herself operating at the level of reality rather than at the level of attitudes toward it. Whereas, in the latter case, attitudes are open to a claim of meaningfulness and truth that potentially permits contradiction, any such contest is superfluous at the level of reality, thanks to its evidential character. The result is "a society that has succumbed to mute nature that has no meaning."[3]

The mute society Kracauer imagines is Facebook society, and its Kracauer is Bernard Stiegler, who describes the falling silent of society as attention deficit disorder and infantilization. The digital media already advance this process with their technological *dispositif* (multitasking, interaction, hyperlinks). The change is accompanied by a shift from deep attention to hyperattention, which, as "hyperstimulation" and "hyperactivity," is associated with additional attentiveness only in the sense of a nonreflective "wakefulness."[4] Stiegler reads this psychogenetic mutation in terms of a theory of power, as a hollowing out of the type of reasoning necessary for every democracy. His cultural pessimism is as unvarnished as critical theory's analysis of mass culture once was, and it has been described as an extension of Hannah Arendt's Eichmann analysis to our entire culture. Stiegler's references for his critique of the present are the emancipatory imperative of the Enlightenment and the ethic of responsibility of the environmental movement. At the core of his critique is the "short-termism" that Stiegler views as embodied in consumerist economics and neoliberal finance capitalism, as "disinvestment" in the future and the establishment of a "society of carelessness." This "short-termism" amounts to the victory of the hedonistic Lovell over the visionary Faust, through which, in Stiegler's view, if nothing changes the world will be destroyed.[5]

The epistemological core of Stiegler's critique is what he calls "cognitive and affective proletarianization": the outsourcing of cognitive and affective matters to technologies and the down-grading of knowledge to information and of experience to know-how. When Stiegler problematizes the uncoded memory of recording media such as videorecorders and computers because—unlike description—they do not actively involve the sender in creating the entity that becomes the bearer of the memory, this links him to Kracauer's media-ontological discussion of photography and hence also to the photography-related view of Facebook society that is offered in this volume. Uncoded memory corresponds both to objects' "spatial appearance," in Kracauer, and to the automatic and automaticized entries on Facebook and visual communication via Snapchat. Stiegler's answer to this loss of cognition is the psycho-technique of writing, as a process of "textualization" that, in describing, analyzing, and resynthesizing the objects under consideration, confers on them a "rational materiality." But the development of social networks, as exemplified by Zuckerberg's "frictionless sharing" and Snapchat's forgettable snapshot communication, points in the opposite direction—toward a visual and indexical materiality that bypasses processes of rationalization. In everyday communication, this shift from thought to materiality is expressed in the move from summarizing to citing, for example when a video or text is no longer explained or summarized to a conversation partner but merely held out for that person to see on a smartphone.[6]

The popular counterpart to Stiegler is Nicholas Carr, who, in his 2010 book *The Shallows: What the Internet Is Doing to Our Brain*, described the internet as an ecosystem of distraction technologies leading to a "switch from reading to power browsing."[7] This, he claimed, renders human action superficial because the surface of the medium, with its link structure, multitasking, and network, works against any deepening of concentration. What other internet researchers and brain scientists praise as stimulation of the brain and as a mode leading to more intensive work is for Carr, along with the brain researcher Maryanne

Wolf (to whom Stiegler also refers), merely a gain in sensory nimbleness that comes at the cost of cognitive acuity. Working online, Carr writes, "requires constant mental coordination and decision making, distracting the brain from the work of interpreting text or other information. . . . We revert to being 'mere decoders of information.'" He fears that the loss of deep reading and deep thinking will also result in a loss of memory, because the objects no longer enjoy sufficient attention to be able to be transferred, via synapse creation, from the hippocampus to the cortex, from short- to long-term memory.

There are four possible reactions to the cultural pessimism of this conclusion. The simplest is agreement. We may argue over how real the danger that has been identified already is and to what extent the interactive culture of digital media aggravates it. But those who join in the warning are on the safe side. How could we not regret the loss of concentrated reading? Who would be ready to declare publicly that we can do without complex thinking?

A second response is to doubt the conclusions arrived at by Carr, Stiegler, and others and to respond to the theory that all this makes us dumber with the theory that, on the contrary, it is all actually producing heightened intelligence. Admittedly, this cultural-optimistic view of the digital media usually rests on a discrete category shift from mental profundity to presence of mind. Claims that the use of digital media leads to more efficient processing of information and greater competency in problem solving harness neuroscientific research to glorify computer games, power browsing, and multitasking as good ways to keep the brain young and active. Whether this argument holds or not, it does not exactly undo the charge that the new media are turning us into mere decoders of information and encouraging a kind of unreflective wakefulness.[8]

The third response is to relativize the cultural-pessimistic finding by hoping that where there is danger, rescue is near—a rescue that can occasionally be found in the very media that are the source of the danger. This is also, in a certain sense, Stiegler's position when he anticipates that the participation culture of digital media will lead to the replacement of the global "mercantile production of memory" with a new era of transindividual memory. The considerations laid out in chapters 1 and 2 cast doubt on this notion. For one thing, the present, precisely thanks to the participatory networks, is no longer really being experienced but merely being transposed so that it becomes a more or less uncoded reception of phatic communication. For another, self-description on social networks is more likely to encourage the episodic model of perception, which lives only in the moment, than the narrative model that serves reflection.[9]

The fourth reaction, finally, is the most problematic: agreement with the conclusion but without the usual negative evaluation. This position does not question the superficiality or antinarrative effects of Facebook but wonders whether this really entails a loss. It defends the episodic model of identity by citing the costs of narrativity—enforced coherence and pressure to respect causality, along with necessarily distorting selection processes—while simultaneously invoking concerns that are more far-reaching than anything mentioned above, such as the construction of ideological systems, cultural identities, and collective memories. In all these areas, namely, narrativity is employed in distinctly problematic ways, reaching from the exclusion and segregation of others to the heteronomy of individuals in thrall to preformed collective memory. Admittedly, once the value of narrative has been called into question, the negative value associated with its endangerment is also up for discussion. New cultural techniques such as hyperattention, episodic identity, and phatic communication begin to seem less like a danger or loss, and the threat posed by such potential future phenomena as a community based on superficiality or the culture of forgetting are no longer necessarily perceived as negative.

The intellectual adventure of the fourth reaction lies in the challenge of uncovering the negative consequences of an essentially positive phenomenon and, vice versa, the positive aspects of a process that is essentially negative. The decisive question is: To what extent are cultural narration and collective memory in conflict with the cosmopolitanism of the Enlightenment and the contemporary debate over human rights? This broader perspective necessarily widens the scope of the previous investigation. We need to inquire, first, into the second aspect of the concept of "Facebook society" before taking another look at Facebook itself as a potential space for the practice of "groundless" togetherness. The starting point is hyperlinks, which are frequently seen, nowadays, as a source of evil but which for many people were once the central point of reference for critical thinking.

SYSTEMS THINKING AND HYPERLINKS

Hyperlinks are the mechanical incarnation of "point-time" and its fractured temporality. The permanent reiteration of arrival and departure that they encourage occupies the opposite end of the spectrum from the principle of continuity, coherence, and contemplation. Some historical distance was required before a critique of hyperlinks began to emerge. When they were new, because they undermined hierarchies and created alternatives, hyperlinks were enthusiastically welcomed as a practice of postmodern theory. Hypertext was celebrated as the "death of the author"—now readers could codetermine the structure of the text! Enthusiasts praised hypertext's structure of networking, reconfiguring, and relativizing, its ability to open up closed texts, as encouraging a constructivist, rather than an objectivist, perceptual perspective. There were those who even foresaw a revolutionary turn toward irony and skepticism.[10]

Twenty-five years later, this nimbus has vanished. Once touted in academic circles as a symbol of critical thinking, hyperlinks are now more likely to symbolize the attack on deep, concentrated reading, as it is constantly interrupted by necessary navigational decisions and the ever-beckoning exit to other communication contexts. The hyperlink has become an antihero—unless we consider it in connection with a thought model that comes from the middle of the last century.

In January 2010, the *Times* (London) likened continuous navigation among different websites to the behavior of a fox. The comparison is based not on the Mozilla browser icon Firefox, which shows a fox embracing the globe, but on the philosopher Isaiah Berlin's essay "The Hedgehog and the Fox," which was written in 1952 and put forward different thought models for the two animals. While people who resemble the hedgehog "relate everything to a single central vision, one system, less or more coherent or articulate, in terms of which they understand, think and feel," foxes pursue contradictory aims and

> lead lives, perform acts and entertain ideas that are centrifugal rather than centripetal; their thought is scattered or diffused, moving on many levels, seizing upon the essence of a vast variety of experiences and objects for what they are in themselves, without, consciously or unconsciously, seeking to fit them into, or exclude them from, any one unchanging, all-embracing, sometimes self-contradictory and incomplete, at times fanatical, unitary inner vision.[11]

In experience and thought, the fox is the episodic type, who doesn't combine different elements into a greater whole but lets them be in their individual distinctiveness. It is the relationship model for Facebook society, since the internet (according to the *Times*, fifty-five years after Berlin) has turned us all into foxes: "We browse and scavenge thoughts and influences, picking up what we want, discarding the rest, collecting, linking, hunting and gathering our information, social life and entertainment."

Following Berlin, a newly positive value is ascribed to hyper-attention in political terms, as well. According to the *Times*, hedgehog thinking is fundamentalist, while the fox's method of thinking threatens totalitarian ideologies; this explains why "the regimes in China and Iran are so afraid of the internet." The conclusion may be a bit hasty—totalitarian regimes don't only fear the lack of ideology; they also fear ideologies that oppose theirs—and it is quickly abandoned. But the starting point corresponds quite well to the media-theoretical assumption that new media change not only the way knowledge is presented and distributed but also the way we treat it more generally. Thus, the printing press, by providing identical, paginated, and relatively affordable editions, made possible an intensive scientific discussion that was not confined to religious topics. And, thus, the internet, with its democratic, grassroots-created, and hypertextually structured publications, encourages knowledge creation "from below," along with a mode of thinking that is searching and nomadic rather than lingering and deepening. Does this mean the internet does away with ideological rigidity? Despite all the criticism it has received, could power browsing, as the material realization of foxlike knowledge management, be not the decline and fall but instead a progressive evolution of culture?

The first step in finding the answer must be to note that, for Berlin, the fox was the better of the two models. The world, Berlin thought, is a place of manifold perspectives and contradictory ideas about value, contradictions that cannot be resolved within an orderly system such as the monist Marxism he was criticizing. A person who stubbornly conceives the world from a single perspective will want to convert others to the same worldview, conceivably also using means that go beyond verbal exchange. Foxes, it can be assumed, are more comfortable with irreconcilable views and values and therefore better prepared than hedgehogs for the underlying conflicts of global, multicultural societies. Yet foxes, too, if they don't want everything to come to naught, must also pull things together in some fashion. The model for this is wit, which one of the most famous German foxes, long before the advent of the internet, placed at the center of his aesthetic and pedagogical theory.

The Frankish poet and cosmopolitan Jean Paul[12] was infamous around 1800 for "grotesquely combining things which have no real connection with each other."[13] Jean Paul elevated the criticisms of his work by the systems philosopher Hegel (whom Berlin counted among the hedgehogs) to the level of a program: connecting disparate things. And also: breaking up things that were connected. This can be seen not only stylistically, in Jean Paul's poetics of interruption and digression, but also substantively, in his vehement critique of the construction of systems. In the name of a single dominant idea, systems of thought— here Jean Paul's view resembled Berlin's—boycott everything that doesn't fit in with this idea. "Finally, a guild of systematizers becomes unable to understand anything (except its lingua franca), including, it follows, every opinion." This was the basis for Jean Paul's advice not to wrap oneself in a specific theoretical construct but to be at home in all and none of them: "Defend your higher poetic freedom against the despotism of every system by studying all systems."[14]

Jean Paul's critique of system making was just as grounded in a philosophy of language as postmodernism is. It also had political consequences, for example when Jean Paul, during the Napoleonic occupation of Germany by France, did not take sides with either the Bonapartists or the nationalists.[15] In his *Levana, or the Doctrine of Education*, Jean Paul devoted a whole chapter to this critique, under the title "Development of Wit."[16] It referred not to punning but to a kind of mental open-mindedness toward things that are outwardly heteroge-neous, for wit is the "disguised priest who copulates every pair" by revealing the sameness hidden beneath diversity and distance. Unlike acumen, which takes the reader by the hand and leads her from Alpha to Omega, sequentially and step by step, wit does not give access to its workings but presents the result as a surprise all the more effective for being unexpected. The "soul" of wit, in Jean Paul's formulation, is its brevity. Its "aha" effect is often an intuitive insight whose trustworthiness must be put to the test by carefully retracing the steps that were omitted. It is for this very reason, because wit as an intellectual practice is an exercise more of thought than of memory, that educating children in wit is so central to Jean Paul's pedagogy.[17]

A particular form of this pedagogy of wit is "learned wit," which makes reference to everything—"all customs, eras, knowledge"—and thus brings different social and geographical circles of knowledge together so as to include, for example, scholars of religion and law, residents of big cities and small towns, trainees and businesspeople. The purpose of bringing all these circles together is profoundly political: "Namely in the end the earth must become *one* country, humanity *one* people." Thus, wit, which, whether "learned" or "coupling," can be likened to the concentric circles of Stoic cosmopolitanism, becomes the central tool of a pedagogical and political utopia. It serves an informational model that transmits knowledge across all borders and beyond all expectations. It is the "little brother" of printing, for books, as Jean Paul noted, also create "a universal republic, a club of nations or a Society of Jesus in the more beautiful sense or human society."[18]

An obituary for Jean Paul described him as having been far ahead of his contemporaries and pictured him waiting at the gate to the twentieth century, "until his laggard people catches up with him." Since then, Jean Paul researchers have been debating his (post)modernity. The 250th anniversary of his birth, in

2013, found him as out of touch with the times as he had been when he was alive. But when it comes to new media, the conclusion should actually be rather different. Hasn't encyclopedic cosmopolitanism taken concrete form in the "Cosmopedia," as the utopia of knowledge society on the internet, which transcends all national borders? Hasn't systems critique become a daily practice, in the form of power browsing? Doesn't the "development of wit" occur in a continuous encounter with hyperlinks? Before we answer these questions, we should explore some related issues more deeply, and so we turn to a German philosopher who may not have been among the "laggards" but who in a certain sense was nevertheless among those who couldn't quite catch up with Jean Paul.[19]

In 1851, Arthur Schopenhauer, in his notes on thinking for oneself (*Selbstdenken*), warned against reading too much, because in the continuing encounter with foreign thoughts the mind does not get around to any formulating of its own:

> To think with one's own head is always to aim at developing a coherent whole—a system, even though it be not a strictly complete one; and nothing hinders this so much as too strong a current of others' thoughts, such as comes of continual reading. These thoughts, springing every one of them from different minds, belonging to different systems, and tinged with different colors, never of themselves flow into an intellectual whole; they never form a unity of knowledge, or insight, or conviction; but, rather, fill the head with a Babylonian confusion of tongues.

Accordingly, the possibility of achieving a coherent system of thought is inversely proportional to the number of influences on its ideas. The critique is directed *avant la lettre* at the foxes' desire to be in many places at the same time. Schopenhauer contrasts "minds which are full of mere antiquarian lore; where shreds of music, as it were, in every key, mingle confusedly, and no fundamental note is heard at all" with those thinkers who, like Berlin's hedgehog, are "strong enough . . . to master [knowledge], to assimilate and incorporate it with the system of [their] thoughts, and so to make it fit in with the organic unity of [their] insight, like the bass in an organ, [which] always dominates everything, and is never drowned by other tones."[20]

The warning about a Babylonian confusion of languages in Schopenhauer's notes on thinking for oneself was by no means original. At the end of the eighteenth century, Johann Gott-fried Herder had already been complaining that, in the "printed Babel" of the world of books, the ideas of all the nations were flowing together and that "innumerable competing foreign thoughts" were endangering the peaceful development of the individual's own ideas. Herder's conclusion in the 1790s shows that information overload was recognized as a problem long before the dawn of the information society:

> And if, every day, only ten daily newspapers and journals fly at you, and in every one only five voices resound in your direction; where, in the end, do you have your head? Where do you have time left for your own reflection and for conducting business? Evidently our printed literature is invested in completely confus-ing the poor human spirit and robbing it of all sobriety, strength, and time for quiet and noble self-cultivation.[21]

Jean Paul also reported that "the book-pollen flying everywhere brings the disadvantage that no people can any longer produce a bed of flowers true and unspotted with foreign colours." But at the same time, the system critic, lover of wit, and conjurer of a universal republic of books also saw in all this an advan-tage, namely in the promise that "through the Ecumenic Council of the book-world, the spirit of a provincial assembly can no longer slavishly enchain its people" and that "the citizen of the world . . . under the supervision of the universal republic, will not sink into the citizen of an injurious state."[22]

At first glance, Jean Paul's and Schopenhauer's perspectives both seem plausible for the diagnosis of Facebook society. For one thing, the internet, even more than book publishing, undermines the sovereignty of every nation-state when it comes to controlling information. For another, modern knowledge management via search engine, copy-and-paste, and hyper-reading nostalgically recalls Schopenhauer's reminder that an insight a person "could have found . . . all ready to hand quite complete in a book and spared himself the trouble . . . is a hundred times more valuable if he has acquired it by thinking it out for himself" because "it is only when we gain our knowledge in this way that it enters as an integral part, a living member into the whole system of our thought."[23] After the revolution in reading that took place around 1800, the history of media, except for the early years of radio, seems to have gone against the hedgehog. "Radio, television and newspapers," observed the Italian philosopher Gianni Vattimo at the end of the twentieth century, have become "elements of a general explosion and proliferation of *Weltanschauungen*" that render "any unilinear view of the world and history impossible."[24]

Since then, the internet has increased the number of voices even further, and it would thus appear that the media, in their development, are putting into practice the very slogans advanced by postmodern philosophy: difference and pluralism. In media society, "the ideal of emancipation modeled on lucid self-consciousness, on the perfect knowledge of one who knows how things stand," begins to be replaced by the ideal of an emancipation based, in principle, "on oscillation and plurality and, ultimately, on the erosion of the 'reality principle.'"[25] Self-skepticism as emancipation, plurality in place of unshakeable conviction—Vattimo's statement clearly puts him closer to Jean Paul than to Schopenhauer. So, is Schopenhauer untimely, and Jean Paul, with his wit, as up to date as Isaiah Berlin with his fox? Does the internet, with its mash-up of the most divergent things, operate in the sense of Jean Paul? Of Berlin? Or Vattimo?

Initially, links may seem like a technical updating of Jean's Paul's wit. But on further investigation, also taking into account the hyperactivity they create, links are revealed as something closer to its demise. Links do not operate like priests or couplers, bringing apparently diverse things together; instead, they operate as duplicators of themselves, busily and noncommittally linking to other links that, in turn, lead to even more links. In multitasking and power browsing, links are subjected to the treatment Jean Paul's wit received at the hands of its female readers: "If they happen upon scholarly wit, they don't cry out rudely, or complain of being disturbed, but rather read on quietly and—the more easily to forgive and forget—do not even want to know what was actually meant." In the era of hypertext and hyperattention, this so-called light reading of "womenfolk"—a "girlish gaze," it is called elsewhere—is the new norm. A link that does not offer up its content to intuitive understanding leads not to a search for the deeper context but to "turning the page." Nor does it come to an indictment, as Jean Paul imagines, in the form of a court battle of readers against his digressive turn of thought, for an indictment of things that are incomprehensible would necessarily assume readers who would go to the trouble of understanding.[26]

The ironic tone of forbearance in Jean Paul's quoted remark on female reading habits becomes more imploring elsewhere: "The flood of books dries up, leaving only a couple of husks, floods your memory once again, and after this ebb and flood there remains in your soul not a single watered plant, but a wet, sandy desert." This is how Jean Paul describes the dementia-like effects of reading too much too quickly. To counteract this, he recommends not Schopenhauer's readerly diet but a second and third reading of the text. Jean Paul's answer was excerpting and indexing. He filled notebooks upon notebooks with curiosities and created indices to provide an overview, in a process that anticipated databases. In this way, he created order in multiplicity—an order, however, that, like databases, does not shy away from the reorganization that necessarily always awaits the person who harvests multiple, diverse fruits of reading far from the comfortable cultural circles of home.

Jean Paul's image of the wet, sandy desert describes the condition of Facebook society more accurately than Berlin's fox. The sandy desert symbolizes the loss of deep reading and deep thinking, for the likes and shares of the hyperactive seek neither to excerpt nor to recombine things but merely to get rid of them. Could this also be the contemporary continuation of the emancipation ideal that, according to Vattimo, characterizes the mass media? Is the sandy desert the expression of radical oscillation and plurality—so radical that nothing that runs through the fingers during the leap from link to link can still manage to take root? Is it, perhaps, just these wet, sandy deserts of people's memory in which the future of the world is growing? The extent to which this question is justified will become apparent in the course of reflecting on the antipluralist aspect of collective memory. Before that, however, it is necessary to shed some light on how memory works and on the capacity of the internet to function as memory, an inquiry that will take us even further back in history, before Berlin, Schopenhauer, and Jean Paul.

MEDIA MEMORY

Once upon a time, the Egyptian god Thoth wanted to give King Thamus the gift of writing. The king refused, arguing that if people could write everything down they would forget how to remember. This is what is written in Plato's *Phaedrus*, which gives a very early explanation of the relationship between memory and forgetting, couched in terms of contemporary media development. The invention of writing is the first caesura in the interrelationship of media and memory, for, with it, memory was no longer tied to individuals who remembered, and storage was no longer a matter of oral transmission. Whereas in oral cultures rhapsodists and priests determined how the past was remembered, written transmission strengthened the position of the past within the present. For when words become separated from the speaker, the dead can also join in the conversation.

This, it is true, can happen only in conformity with the requirements of the living. Written material is always picked up and communicated by actual individuals in concrete situations. If the past is externally stored, more can be retained than any one person can convey to another or would wish to. This is why research on memory distinguishes between storage memory and functional memory. The data that are needed—to use an analogy with computers—are fetched from the depths of society's hard drive (past tradition) and transferred to its random access memory, or RAM (present communication). Society's spokespersons and discourse leaders determine which data are required and permit only those things to rise up to the collective memory of the present that correspond to the politically desired version of the past. Undesirable memory material does not form part of the desired tradition and is suppressed, until, under the leadership of new spokespersons, its hour comes. For an important part of strategic remembering (for example, acts of nation founding and their heroes) is strategic forgetting (for example, the violent deeds that may be associated with the events in question). While storage memory—the archive— contains what *can* be said, functional memory determines what *is* said.[27]

The selective mobilization of stored material transforms the "passionless archive" into an "emphatic site of memory" that can be used to create collective meaning.[28] In the context of the present discussion, a media-theoretical parallel can be drawn. Namely, the archive of storage memory relates to the functional memory of things currently being recalled the way photography is related to painting. While photography takes in whatever happens to be in front of the camera lens, painting records only what happened (somehow) to be (in some sense) in the consciousness of the painter (and hence was important to him). Thus, for Kracauer, photography's inventory-like quality—as "barren self-presentation of spatial and temporal elements"—corresponds to historicism, which was related by Nietzsche to the third, "antiquarian" type of relationship to the past: a "blind mania for collecting," pedantic and passionless, that salvages the past for its own sake, while "monumental" and "critical" relationships to the past both treat the latter as being, respectively, emphatically either positive or negative.[29]

Historicism's passionless mania for collecting corresponds to the archivist's passion for preserving things, with an emphasis on their registration. The ideal archivist, typically, is not interested in a meaning-creating history in which diverse data assume the role of evidence. Archivists are not storytellers—out of respect for the material. Narrative takes place outside the archive, in the media, in schools and universities, on monuments and days of commemoration. The emphatic memory sites of a nation are the sites of "exosocialization" that serve to construct and communicate national identity through the corresponding creation of historical events and national myths. This is where the national biography is written. Like autobiographical narratives, it provides orientation in the present by retroactively constructing the meaning of the past.[30]

Even if archivists are more like "photographers" than "storytellers," the archive is not a photograph of the world. It cannot store everything that happens in the world, and it is no less determined by its choice of specific perspectives toward the world than photography is. Unless, that is, the world itself is taking place in an archive. Increasingly, this is precisely the case.

The most recent caesura in the interrelationship of media and memory is the internet, which, with its social networks, invites people, on the one hand, to communicate their private information, while its search engines make it possible, on the other hand, to open up the archive that is created in this way—at least to the extent that everything happens this side of the dark net. Since, in principle, everything that exists in digital form is archived and can be accessed, the internet signifies the end of forgetting. There is still a life outside the internet, but on the internet there is (almost) no life left outside the archive. If photography was "the *general inventory* of a nature that cannot be further reduced,"[31] then the internet is the general inventory of digitally represented society. The internet of things, which lets our cars, items of clothing, refrigerator, coffee machine, radio, heating system, lights, etc., talk to one another, expands the terrain invaded by inventory by transforming even the objects of daily life into small, powerfully effective archives that all tend toward the creation of one gigantic central archive. The internet—not in the way it is used but in its content—restores to the archive the innocence it had lost, at least since Michel Foucault's critical works on the archaeology of knowledge and the genealogy of power. The foreseeable future is the archiving of the entirety of existence, complete with all its everyday and less history-worthy details—a kind of 1:1 map that, unlike the 1:1 map in Jorge Luis Borges's story "On Exactitude in Science" or, before that, in Lewis Carroll's novel *Sylvie and Bruno*, is actually quite useful thanks to search engines and algorithms.

With the internet, and especially since the Web 2.0, the problem is no longer to be found in the lack of capacity to remember but rather in the disappearing possibility to forget.[32] However much we may like to live in the moment (or, in our eagerness to communicate, may in fact not even properly experience it), and however little we may experience ourselves as part of a history (a history that makes some kind of sense and has something in mind for us), technically every Now becomes an unlost past. Every casual communication on social networks continues to be stored somewhere and fills the data pool from which sociologists, marketing specialists, and secret services look forward to deeper knowledge of society.

The increased storage function of the internet should not be seen only in relation to these new technical possibilities, however. It is fundamentally the technological radicalization of a social trend that was already diagnosed for the 1980s: a massively expanding mania for archiving, accompanied by an explosion of discourse on memory. While the twentieth century began with visions of the future that, after the socialist revolution in Russia, contributed to a fear-inducing realpolitik, it ended with an obsession with the past. The fact that the temporal focus has shifted "from present futures to present pasts" can be explained by the loss of hope in the future but also by a vanishing familiarity with the present, as the acceleration of social processes leads to a "shortened stay in the present" and as the end of grand narratives contributes to a "culture of memory."[33]

The culture of memory and the cult of archiving may spring from the same causes, but when it comes to their goals, they are not only different but quite contradictory. While discourses of memory aim at a narrative (re)ordering of the past—brought about, among other things, by the round-number anniversaries, during the 1980s, of events that occurred under National Socialism as well as by feminist and postcolonial critiques of previous images of the past—the obsession with the archive sidesteps narrative order in its turn toward the facts as such. As the French historian Pierre Nora describes the situation: "Memory has been wholly absorbed by its meticulous reconstitution. Its new responsibility is to record; delegating to the archive the duty of remembering." The construction of history gives way to the archiving of history, while "the emphatic site of memory" is replaced by the "passionless archive," which remains neutral in terms of content and the impulse to remember.[34]

The shift from the narrative structuring of events to their indifferent registration recalls Jean Paul's "wet, sandy desert" where unprocessed information lies fallow. At the same time, the "outsourcing" of the unremembered past to the archive reenacts the previously discussed shift from long-term experience (*Erfahrung*), as meaning-conferring *interpretation*, to short-term experience (*Erlebnis*), as distanced *information*, inasmuch as, once again, the individual fails to make the perception fully "her own." The basis for this shift is the gradual suppression, in the course of media development, of the *"Aufschreibesystem"* (notation system) writing, which, as a mode of description, is inevitably subjective and meaning conferring, and its replacement by recording techniques like photography and (video)recording, with their objective registration. The mechanical reproduction of reality, which became a mass phenomenon with the availability of inexpensive cameras at the end of the nineteenth century, expanded audiovisually during the twentieth century. In the twenty-first century, thanks to digital media, it took a further qualitative leap in terms of its reach, extent, and analytical possibilities. The "self-musealization per video recorder" in the 1980s now appears to have been the modest prelude to the twenty-first century's permanent and automatic self-archiving on personal websites, social networks, and diverse self-tracking apps. This self-representation aims, as was demonstrated in chapter 2, less at narrative self-understanding than at the creation of an archive of the self in the mode of recording.[35]

Having established the distinction between archiving of the past and memory of the past, we can now give the statement that with the internet the possibility of forgetting is shrinking a sharper dialectical focus. The internet is the end of forgetting, to the extent that it potentially archives everything, but at the same time it is also the end of memory, to the extent that the unlost past is no longer a formed past at all. The stored information lacks the structure of narration and the perspective of the storyteller. The internet is not "memory in an emphatic sense," which would keep the past meaningfully present; it is a "radical presence (or latency) of data in storage." This is why it is seen by some not as an intensification but rather as a "withdrawal" of (cultural) memory: "The World Wide Web, as an apparently navigable archive, claims memory, but practices its opposite: amnemic [sic] rituals of cybernetics." The internet, through its ideology of "contentism," basically creates "digital amnesia." It is—the distinction is crucial here—not a memory that has processed experience but merely an archive that has deposited information.[36] What does this mean?

When remembering—rather than forgetting—is declared to be a void on the internet, the concern is not with its archival subfunction from the perspective of a materially oriented media theory. Naturally, access to stored material depends on the present availability of past storage techniques. The format of the bearers of saved material itself becomes something worth saving, as everyone who, in 2017, tries to open a Windows 98 file or load a .gif file from a floppy disk knows. Digital media have a "memory Darwinism" at their core, according to which only things that are continuously utilized (and thus constantly demonstrate their significance) remain in storage (when it is updated to the contemporary format). The problem here is not the obsolescence of soft- and hardware but the insecure status of the medium itself, starting from the fact that every individual who has or can obtain access to the server is able to manipulate the stored material. While analog writing on solid surfaces enables the presence of the dead in the discourse of the living, digital writing on the internet allows the present to secure a place in the discourse of the past. More decisive than possible human intervention into the archival material, however, is the media-specific response of the archive to the individual who is accessing it.

The unreliability of the internet as memory is inherent in its paradoxical nature as an individualized mass medium. In the context of our current discussion, this is less a matter of the possibility for potentially all individuals to address themselves to potentially all other individuals than it is a matter of the customization of the content, in each case, for the specific user by means of cookies, browsing histories, login data, and other forms of identification. While conventional media—books, newspapers, radio—provide all recipients with the same news, the internet adapts the news to the recipient. Processing is personalization, as anyone knows who, for example, compares the results of a given search with the results obtained by a different person who poses the same question at a different place and time. In the course of this "mass customization," the search engine produces "memories that have admittedly never been thought before and are merely the product of the context-related commands of the user."[37]

The dynamically produced information already represents a turning away from collective memory, for it addresses the individual not as part of a group but in her particularity. The questions an individual asks of society's memory—Where do I come from? Where do I belong? Who are we?—are preceded by the trace-following algorithm: Who is this user? What websites has she previously visited? When the individual becomes the pretext and context for shaping the information provided, her specificity determines the current presentation of collective memory rather than the materials of collective memory affording the context for the individual. Cookies are the negation of collective memory inasmuch as they gather data on an individual rather than a collective basis and distribute information (including collective memory) on that same basis. In this case, the individual's behavior does elicit the information, but it does so unconsciously and only via computer-generated surveillance.[38]

The technical disposition of the digital media is not only detrimental to collective memory by virtue of the individualizing function that cookies perform. Hypertext, as a technology of necessarily individual text montage, already poses a challenge to collective memory. For as soon as the elements of a text can be mounted in varying ways, the principle of invariance and replicability, as the fundamental model for memory, is endangered. Whereas writing, as a technique, replicated reality in ways that were more standardized than the previous oral tradition, hypertext leaves it once again open to individual variation. Basically, this trend toward individualization is a tendency that is already embedded in computer functionality, since the computer is a machine with no set determination, which can be utilized for the most diverse purposes—as an adding machine, typewriter, reading tool, television, or supermarket—and always operates in interaction with its environment. Thus, when it comes to information transfer, the computer also provides a user-oriented selection process incorporating heterogeneous components in which cookies, hypertext, search engines, and dynamically constructed websites are both the means and the result. Since the Gutenberg galaxy, collective memory has lived on in individual objects that serve as their bearers (printed books) while varying with the individually specific mixture of these objects based on an individual person's reading. Now, in the era of digital media, the provision of the material of memory to the entities that serve as its bearers (individual computer screens) is already variable.

* * *

Memory, to sum up, is shaped both by specific social forces that determine the content of the archives and also by the media employed for this purpose, which bring specific techniques of remembering with them. The past is not a construction that is only cultural. It is also media dependent, and media are not neutral bearers and conveyors of the content of memory—they also shape the modalities of remembering. Whereas writing secures the material of memory against the variability and loss entailed by personal reproduction in an oral culture, and printed books multiply the reliable replication of the past, photography and video, with their optically unconscious fidelity to detail, also recall what was *not* perceived in the past. The internet, finally, potentially archives everything that is presented on it. To the extent that this archiving occurs without selection by experts or intentionally appointed persons, it can be understood as a democratization of the archive. To the extent that it is all-embracing, in a certain respect it signifies the end of forgetting.

The cost of this end of forgetting is the relinquishment of recollection. Memory and recollection are meaning-creating (narrative) processes for structuring archival material. The selection processes they use are constitutive of (collective) memory and are binding for their addressees. This very practice of collective remembering is undermined by the technical *dispositif* of the internet: by hypertext, process dependency, and personalization. The internet is not detrimental to collective memory because it would be unable to secure the archiving of the material. On the contrary, the deterritorialized, participatory nature of the internet even makes possible new forms of recollection while simultaneously rendering potentially repressed memories globally accessible. However, these new possibilities of collective remembering only exist within a technological framework that is detrimental to collective recollection because the processing of the material, for the reasons previously mentioned, must necessarily be unstable. At the concrete level of the content of a static website, this may not be noticeable, but at the general level of information accessibility on the internet, we have to say: "The digital engineering of collective memory is no longer a function of social filters but of programming." The canon of collective knowledge, which has traditionally been passed down indiscriminately to the members of a cultural or national community via sites of exosocialization, is now dissolving into the computer-generated personalization of knowledge.[39]

The uncertain future of collective memory, though, appears less threatening in view of the problematic consequences entailed by this type of memory. Collective memory, as the orientation and construction of meaning, is always also a normative straightjacket that contains readymade conceptions of value and predispositions to action while more or less unforgivingly pursuing deviance. To put it pointedly: "To claim the right to memory is, at bottom, to call for justice. In the effects it has had, however, it has often become a call to murder."[40] In this context, a "culture of forgetting" could seem attractive. Let us take a closer look at the negative sides of collective memory and of the cultural narration connected to it, paying attention both to the relationship between memory and narration and to the relationship between identity and cosmopolitanism. In this exploration, the new media will retreat into the background for awhile, only to return in the final section of the chapter, where Facebook is presented as the site of a global community bereft of memory, narration, and identity.[41]

NARRATION AND ENFORCED COLLECTIVITY

The critique of the paradigm of narration began, in chapter 2, with the defense of the episodic personality type against its narrative corollary, the turn of postmodern poetics against the illusion of coherence, and the view of automated biography as a protection against techniques of narrative self-deception. Now, it is necessary to complement this perspective, drawn from literature and autobiography, by looking at its political dimension, in light of the fact that narrative self-perception always takes place in a realm influenced by cultural and social symbolic systems. These symbolic systems overlie individual perception and narrations of reality and inevitably lead to discrimination, exclusion, and distortions that favor a coherent identity and perfect identification. The accusations made against cultural narration and collective memory are essentially three: perceptual distortion, enforced coherence, and heteronomy.

Heteronomy comes into play when success in life is bound up with the success of an individual's narrative as a meaningful, goal-directed story that conceives the I as part of an other-directed project: as the child of specific parents, the citizen of a specific city, the member of a specific professional grouping, clan, tribe, or nation. "Identity, like memory, is a kind of *duty*" is how Nora puts it, continuing: "I am asked to become what I am: a Corsican, a Jew, a worker, an Algerian, a Black. It is at this level of obligation that the decisive tie is formed between memory and social identity."[42] The little history of the individual is subordinated to the big history of the collective and is determined by the latter's culture and memory. Doubt about the ethical force of narrative is thus expanded to encompass a critique of the narrative form of self-observation as a means of social discipline—as the instrumentalization of the Now under the sign of a future projected forward by the past of others. The alternative is to be found, accordingly, not in a counternarrative but in narrationless snapshots. This points us back to Facebook's episodic status updates.

Even before stories and their relational content, the heteronomy, or other-directedness, that narration imposes on the individual begins at a formal level, with the requirement of coherence. The "seamlessness" expected of narration suppresses the breaks and contradictions of individual life histories and identity concepts and may thus obscure the truth precisely in the interest of presenting a formally anchored illusion. Narration is then not only not the better form of self-understanding but possibly even the worse one. As Judith Butler puts it in *Giving an Account of Oneself*, "if we require that someone ... be a coherent autobiographer, we may be preferring the seamlessness of the story to something we might tentatively call the truth of the person." The ultimate consequence of this perspective is a suspicion that ethical responsibility may lie precisely in the "non-narrativizable exposure that establishes my singularity."[43]

Both of the points made by this critique—the content-related one of narration's goal-directedness and the formal one of its coherence—need to be discussed not only in the context of individual identity formation but, more broadly, at the level of culture(s). Cultural "metanarratives," to employ Lyotard's concept, are also continuously constructing coherence and suppressing alternate perspectives. Individuals are compelled (or may themselves seek) to evaluate actions and interactions from an external narrative and to perceive their "daily narratives" only as elements in a "unified narrative":

> There are second-order narratives entailing a certain normative attitude toward accounts of first-order deeds. What we call "culture" is the horizon formed by these evaluative stances, through which the infinite chain of space-time sequences is demarcated into "good" and "bad," "holy" and "profane," "pure" and "impure." Cultures are formed through binaries because human beings live in an evaluative universe.

The goal is to homogenize difference, in the illusion of a secure identity that is then advanced in opposition to other cultural identities. Moral and political autonomy—this is where the

political weight of the critique of narration is to be found—
exist only through the possibility of acting *outside* the framework of the cultural narrative, including the option of bringing
opposing narratives and loyalties together as one.[44]

The tension between "first-order deeds" and "second-order
narratives" is comparable to the tension between event and
narrative in historiography and also to that between archives
and memory in the discourse on recollection. The difference, in
each case, is one of ownership. As an event, the phenomenon
belongs to all; as an element in a narration, to those who are
telling the story. "Memory is life, borne by living societies
founded in its name," Nora writes, while history "belongs to
everyone and to no one, whence its claim to universal authority."[45] The contradiction behind this tension between event and
narration, individual and culture, is ultimately that between
universalism and particularism, in which something unbound
and unique (an event, an individual) winds up closer to the universal than to the particular. This perspective moves our theme
nearer to politics and the discussion of universal human rights
as a corrective to standards of cultural identity. A productive
starting point for this reflection is the Enlightenment's debate
on cosmopolitanism, which followed the debate on the Babylonian confusion of languages considered earlier in this chapter.

A symptomatic example of the discussion of universalism and
nature versus culture and the nation is found in the Freemason
conversations *Ernst und Falk*, written by Gotthold Ephraim
Lessing in 1778. On the subject of world citizens, Lessing noted:

> If a German meets a Frenchman at present, or a Frenchman an
> Englishman, or *vice versa*, then it is no longer a mere man meet-
> ing a mere man, who by virtue of their identical nature will be
> attracted one to the other; but a particular kind of man meets a
> particular kind of man, who are conscious of their different

tendency, which makes them cold, reserved, suspicious of each other, even before they have the slightest dealings with one another for their individual selves.

The *mere* human being who is addressed here, the human being per se, is the human being beyond cultural memory and social-political reality, as opposed to the concrete human being within a social context. The unavoidable cultural identity of humans, as Ernst and Falk ultimately also agree, seems to be an anticosmopolitan fact of life.[46]

The utopia of universalism barely survived the end of the eighteenth century, and the nineteenth century became a century of nationalism. In Germany, this also took the form of a theory of education: Johann Gottlieb Fichte's *Addresses to the German Nation*. Fichte's addresses, in contradiction to Jean Paul's treatise on education only a few years previously, by no means call for the exercise of wit or for systems critique. Instead, they call for "creation of that supersensuous world order in which nothing becomes, and which never has become, but which simply is forever."[47] It is the revocation of Jean Paul's (post)modernism of in-betweenness and becoming. Education, in the nineteenth century, had become a place for training people to be German, French, or English. As a site of exosocialization, the education system shapes collective memory. This is where national narratives are passed on and put to the test; it is where the nation is created as a narrative. In the twentieth century, nation-states, which had assured themselves of their own identity by means of exclusion, carried out two devastating world wars. In the twenty-first, the human rights movement is salvaging the utopia of universalism that was entertained by the Enlightenment and is carrying it forward as "a new universal language" for strengthening the "unbounded universal 'we'" of humanity against the "bounded 'we'" of nations and cultures.[48]

If, around 1800, the line of conflict was drawn between universalism and nationalism, now it falls between human rights and multiculturalism. While multiculturalism strengthens the untouchability of the given culture, universalism emphasizes the rights of the individual over those of the group. This position rests on central characteristics of the Enlightenment, modernism, and postmodernism that can by no means be assumed or required of numerous other cultures. Therefore, according to a justified objection, the conviction that individual rights rank above collective goals is nothing but "a particularism masquerading as the universal":

> For mainstream Islam, there is no question of separating politics and religion the way we have come to expect in Western liberal society. Liberalism is not a possible meeting ground for all cultures, but is the political expression of one range of cultures, and quite incompatible with other ranges.[49]

In other words, the *mere* human being that the liberal, culturally indifferent perspective aims at is always already a particular kind of—liberal—human being. The expression of this contradiction is not confined to tensions with Islamic fundamentalism or disagreement over the freedom to publish caricatures.

The defense of universal human rights against the pluralism of multiculturalism is absolutely necessary as a protection for individuals, including women, against their cultures' repressive requirements of conformity. At the same time, the affirmation of universal values is not unproblematic, because, in a turn away from the postmodern position of inescapable difference, it assumes absolute rights outside of concrete contexts for action or understanding and thus restores the belief in eternal values.[50] Whatever position one takes in this debate, it is evident that the liberation of the individual is, at the same time, an attack on the collective memory of the culture to which the individual "belongs," or into which she is born, and which now functionalizes her life as a part of its "project." The possibility of absolutely disarticulating these things must be considered doubtful, for the individual only becomes a subject through language, and it is through language that the subject is simultaneously addressed and constituted. There is no articulation of the self by which a subject could escape this prehistory of a "we." In Butler's pointed formulation: "I always arrive too late at myself."[51]

Nevertheless, we can only conclude that the more individual and impervious to the influence of collective narration the subject understands herself to be, and the more she expresses this, the closer she moves to the position of the human being *as such*. This human being *as such* should, however, not be conceived as empty and abstract but also, simultaneously, as *such a one*, possessing her incomparable individuality (or mix of identifications), which transcends any cultural or national categorization (German or English, Muslim or Christian). In this sense, the human rights debate operates on a level both above and below collective identities, with the goal, if not of dissolving, then at least of upending the hierarchy: First, a human being is a human being *as such* (per se and for herself) and only then *such a one*, that is, part of (and in the service of) a community. This change in emphasis implies another—critical, skeptical—look at the authority of the cultural narratives the person has grown up with. It leads, as Lessing, once again, made clear, to conflict with the fathers.

In his play *Nathan the Wise*, Lessing makes no distinction among Christians, Jews, and Muslims but treats them all simply as human beings. With this, he offered up a more optimistic model of society than he had in *Ernst und Falk*, where the differences between Germans, French, and English appeared unbridgeable. The change may be a result of the altered perspective, for now Lessing is discussing the abstraction of humans from their concrete environment in terms of religion, not national and territorial identity, and is doing so in the form of a story, not a philosophical dialogue. It is a narrative, if you will, about the way postmodernity deals with old metanarratives.

The play takes place in twelfth-century Jerusalem, where Recha, the foster daughter of the Jew Nathan, is rescued from the flames of her burning house by a Christian Knight Templar. One day, the Muslim ruler Sultan Saladin, who needs money, asks the rich merchant Nathan to tell him which of the three great religions is best. Nathan answers with a parable about a father with a miraculous ring that he turns out, after his death, to have given to each of his three sons. The judge called in by the sons refuses to decide which ring is the genuine one and proposes a competition without a finish line, arguing that if the ring really makes its wearer beloved among men, the owner is probably the one who is loved by the most people. Since this question cannot be answered by the brothers alone, the judge orders the "children's children's children" of the parties to reappear before his bench in "a thousand thousand years" (3.7).

Nathan, with the ruse of Scheherazade, rescues himself from Saladin's trick question by telling a story. But Saladin refuses to accept this flight into aesthetic fancy and presses for a clearer answer, whereupon Nathan explains the interrelationship among truth, memory, and cultural identity:

For do they not all ground on history?
That's written or traditional? —And is
Not history the only thing which must
Be taken on good faith? —Or is it not so?
Well whose good faith then is it? Which the least
We doubt? Is it not that of our own people?
Not theirs? Whose blood we are? Not theirs, who from
Our infancy did prove to us their love?

Who ne'er deceived us, only, where to be
Deceived was the more wholesome thing to us?
How can I give less credit to our fathers
Than thou givest thine? Or on the contrary. . . . [52]

The commonly held view, in religious circles, that a person can only be truly tolerant if he is firmly convinced of his own belief is not one to which Lessing's *Nathan* adheres. Instead, Nathan's insight into the partisan and contingent nature of a person's own convictions points forward to the skepticism and irony that postmodern philosophy will later recommend as the adequate perspective when it comes to questions of truth and falsehood. This postmodern position *avant la lettre* is a point of criticism that was leveled at Lessing again and again in the reception history of *Nathan the Wise*. People not only disliked the view that a human being is a human first and only then a Christian, a Jew, or a Moslem; the critics also disputed that Lessing

> by tolerance had meant the democratic pluralism that today rules the academic as well as the political scene, that suspects every insistence on long-held truth of being dogmatism, and that promotes the free competition of different views within the community of citizens or of academics as the ideal. [53]

We should not be surprised when a critique of this sort ultimately suggests that the "multitude of contradictory opinions" resembles "a big collection of products," which "appears for its own reasons and offers something for everyone." In more recent theatrical productions, this critique becomes all the more obvious when Nathan, the merchant, appears in a designer suit, thus showing a lack of seriousness toward the expression of personal convictions and demonstrating that tolerating multiple opinions is symbolic of market logic and a consumerist model. The situation appears less clear if one views the relationship between consumerism and ideology in a way that transcends the usual explanations and understands the consumer model as a kind of "pragmatic cosmopolitanism" and as "global society's immune system against the virus of fanatical religions." At this point, if not sooner, the notion of tolerance butts up against its internal paradox, which actually consists *not* in elevating the human being *as such* over people *such as they are* but, instead, in using the model of consumption as a cure for the model of religious conviction.[54]

Lessing's play ends in a "general embrace of all," for Recha, the Knight Templar, and Saladin turn out to be siblings, and uncle and niece. Two figures are excluded from this embrace: Nathan himself and Dajah, Recha's lady's companion. While Nathan's place at one side of the stage remains ambivalent (his efforts to make "all men brothers" having been radically subverted by the metaphor of actual blood relationships), Dajah's exclusion (she does not appear at all in the final scene) is unambiguous. Dajah, as Lessing has Recha remark, is "one / Of those enthusiasts who think they know / The universal, only truthful path / That leads to God!"[55] Dajah cannot accept Nathan's imponderability clause on the evaluation of religions or, along with it, in today's terms, the abstract, coldhearted, elitist, and imperialist tolerance model of cosmopolitanism. With her absolute strength of religious belief and "weaponized" identity, she represents the majority of people—not only in Lessing's time.[56]

Our look at the psychological and political problematics of narrativity leads to the conclusion that there is a dual heteronomy of the subject, which consists, on the one hand, in narrative structure's compulsion to be coherent and, on the other hand, in the readily available content of collective narrations that either offer themselves to the subject as a "home base" for its thoughts and actions or are imposed in the form of tradition. Human rights discourse, in contrast, treats the subject as outside any collective narrative or memory and hence as having recourse to Enlightenment ideas. To the extent that the concept of universal human rights is based on a narrative (the narrative of Western liberal individualism), the only solution that may—possibly—allow the individual to find the free space she requires is negotiating among the various different narratives. This negotiation, one can argue, is more successful the more unconsciously it takes place. The more the mixing of narratives is not just the normative telos of a theoretical concept but also the practical outgrowth of a concretely lived life, the more sustainably it appears to be anchored. Examples of this un- or semiconscious, "factual" cosmopolitanism are the "cosmopolitanism from below" found in multicultural metropolises and the "banal," everyday cosmopolitanism that is an outgrowth of global economic, cultural, and communication networks.[57]

In an era of globalization, when the significance of the nation-state is shrinking and attention to the cosmopolitan aspects of everyday life seems to be a methodological prerequisite of any social analysis, it is only natural if the digital media also (and above all) play a central role. As the most advanced media, they fundamentally define the psychological configuration of the present. And if, as Jean Paul's "universal republic" of books and Vattimo's praise of pluralism in radio and television demonstrate, print and electronic media were already spaces for negotiating the universal, it would seem that the internet, operating beyond the control of nation-states, is advancing the mixing of *we* and *they*, *here* and *there* more powerfully than ever. This supposition is understandable in view of the medium's superregional, intercultural networking possibilities but overhasty if it mistakes the potential for cosmopolitical information processing for its reality. For while national media are as cosmopolitan and multiperspectival as the self-understanding of the country and state in which they operate, the stateless internet is only as cosmopolitical and multiperspectival as its users. Without the oversight of democratic institutions, the individual filter bubble can shut out unwanted information and unpopular perspectives more effectively than ever before. In this process, the easier creation of homogeneous interest groups and the faster, more superficial mode of communication by no means result in increased sensitivity to the other, unfamiliar.[58]

The cosmopolitan impact of the internet very likely lies elsewhere and takes a different form—that of phatic communication on social media. This claim is easier to comprehend if we approach it more psychologically than politically and follow the approach taken by the communication theorist Vilém Flusser, who was born in Prague and emigrated to Brazil in 1940. In his characterization of patriotism, Flusser points to the etymological context of "habit" as "habitation," as a dwelling in which individuals have made themselves at home and feel safe and secure. This perspective gives the usual characterization of patriotism as warmth and comfort, in contrast with the abstraction and coolness of cosmopolitanism, an information-theoretical valence that also has an aesthetic intent: "The noises that approach the dwelling are ugly, because they disturb the habitual. If one transforms them into information, they become beautiful, because they are then built into the dwelling." Patriotism, which is often associated with passion and which in many instances is primarily pride and *amour propre*, is for Flusser the "symptom of an aesthetic sickness" that mistakenly understands what is familiar and comfortable as beauty. This metaphorical approach makes it possible to recognize the theme of difference and tolerance as a problem of information processing that transcends religious, national, cultural, and ideological specificity and to open up new ways of approaching it.[59]

The remarkable thing about Flusser's perspective is his claim that noise that has been processed into information is automatically beautiful. This may appear intuitively plausible, since information says something, while mere noise eludes understanding. However, if we give the aesthetic perspective a semantic turn, it becomes clear that only noise that has been processed into a statement can take a position vis-à-vis other statements and, among other things, is potentially able to contradict them. Information is not only a gain in knowledge and a broadening of the capacity to assign value; it is also, potentially, an experience of difference that destabilizes habits of valuation. To borrow the language of the opening quotation in a way that also challenges it: At the level of the models of values and orientations that we experience as "at home," disturbance of the habitual first takes concrete form when the cognitively foreign (the unknown) is recognized as normatively foreign (as a contradiction); in other words, when noise is transformed into information.[60] Flusser's overhasty talk about the beauty of information per se is only comprehensible in the context of the normative cosmopolitanism on which it is based, which bravely regards every gain in knowledge as enrichment, even when a person's own system of thought is being called into question. Here, we have a communication utopia that recalls Jean Paul's wit and Berlin's fox and that conceives "dwelling"—homeland, identity—not statically, as being, but dynamically, as becoming. Consequentially, Flusser then also sees the real responsibility of humanness as being underway, in a nomadic removal from everything familiar. The migrant, whom the "loss of the original, dimly felt secret of homeland . . . has opened to the secret of being with others," thus becomes its missionary, as the bearer of the "awakened consciousness of all those who have homes, and a harbinger of the future": making a home in homelessness.[61]

Flusser's communication utopia is an actualization of "elite cosmopolitanism," which is set out as a noble goal, not lived as a banal practice. Still, the passage previously cited is central to the discussion that we need to have here, for it contains—let us provisionally assert—the catchword for the *factual* cosmopolitanism of Facebook society. Perhaps the basis of communication across borders lies not in negotiation but in an ignorance of opposing positions. Perhaps the mutual acceptance on social networks results from a connection that, in the phatic mode, never actually takes account of the other as Other. Perhaps the unreflecting cosmopolitanization on the internet consists precisely not in translating noise into information but in enjoying it as noise, or chatter. The proposal may appear absurd (and is not supported by Flusser's communication theory), but in a certain sense it is the recipe for a philosophical theory that, in the final years of the twentieth century, asked us to think a communality that transcends cultural, religious, or political narratives and identities. With this theory in mind, and with reference to some of what was considered in chapter 2, we should now, after the discussion of the relationship among culture, narration, and identity, ask what the social networks' model of phatic communication and episodic self-presentation contributes to an identity formation outside collective narratives. The answer, sensibly enough, begins with the question of the extent to which social networks themselves generate a cosmopolitan narration.

GROUNDLESS COMMUNITY

In 1997, under the title "Birth of a Digital Nation," an article in the Californian technology magazine *Wired* described the emergence of a powerful new form of community on the internet: "young, educated, affluent . . . libertarian, materialistic, tolerant, rational, technologically adept, disconnected from conventional political organizations." Tolerance is listed as one of the fundamental characteristics of these postpolitical and self-referential "citizens of the Digital Nation," as the result of a generous indifference toward specific cultural values and individual forms of life:

> They don't merely embrace tolerance as an ideal; they are inherently tolerant. Theirs is the first generation for whom pluralism and diversity are neither controversial nor unusual. This group couldn't care less whether families take the traditional form or have two moms or two dads.[62]

This notion of an "internet nation" was also entertained in more academic writing predicting that "the netizen might be the formative figure in a new kind of political relation, one that shares allegiance to the nation with allegiance to the internet and to the planetary political spaces it inaugurates." The rationale behind this assumption was the well-established claim that a medium is not simply a tool but carries its own message, able to promote "deep cultural and social changes."[63] The assumption has survived into the present, as demonstrated by the ninth annual meeting of the Internet Governance Forum, in 2014, which held: "Clearly, the Internet provides the basis for a community with its own interests, an incipient identity, its own norms and modes of living together."[64] But is it still possible to hold on to the notion of a nation as media usage at a time when this nation cannot, by any means, still claim exclusive usage of the medium by a technical avant-garde? Can enthusiasm for a technology cover over differences of a religious, cultural, or political nature? Is the talk of the CEOs of Facebook, Twitter, Narrative Clip, and other platforms about "our community" anything more than rhetorical boosting of a business idea?

Reference to a shared currency of the digital community (bitcoin) isn't sufficient to answer the question, and even a consolidated internet time, independent of local time zones, would not suffice as evidence that technology can create a feeling of community.[65] Even a digital nation is not constituted by shared units of measurement but by shared values, behaviors, and information sources. These, then, will have to be sought (in a way that recalls the cosmopolitan approach) beyond the cultures of geographically and politically framed communities. In the case of cosmopolitanism, this abstraction gets caught up in the contradiction that the ideal of a universal human rights regime beyond specific cultures is by no means shared by all people. In the case of virtual communities, whose identity is considered no less "thin" than that of universalist communities, the questions arise: What communicative contents create the feeling of togetherness? In what does the narration of the "digital nation" consist, and what can be said about its memory?[66]

One of the theories about the internet goes as follows: "Where natural social collectives build connectivity out of memory, virtual communities build memory out of connectivity." The conclusion sounds convincing for a "thick virtual community" such as the WELL, which Howard Rheingold described more than two decades ago in his book *The Virtual Community*.[67] Today, too, it is possible that smaller, exclusive virtual communities, whose feeling of belonging is based on shared values, losses, and hopes, may still create a collective memory out of their connection on the internet. But it is doubtful whether the wealth of updates on Facebook shapes stories that can be retold in such a way as to create the collective memory of a "Facebook community." Rather, we may suppose that the basis of the new feeling of community is to be found in phatic communication: beyond meaningful contents and points of reference, in the communication *as such*, which recalls nothing but its own doings. This amounts to a reformulation of the theory mentioned previously: Virtual communities create neither connectedness based on recollection nor recollection based on connectedness. Instead, they constitute themselves in the mode of shared forgetting.

Up to this point, the discussion suggests that we not think of forgetting as a lack but rather see the absence of concrete cultural definition as the possibility of a cosmopolitical opening. This proposal is by no means without precedent. A 2002 essay on Europe's postmodern identity declared that the only way for the various peoples and cultures of the European Union to develop a common feeling of belonging was not via an emphasis on collective remembering but through a collective loss of memory that would aim to forget, together, the centuries of war and conflict of Europe's peoples and nations.[68] The proposal recalls notions of reconciliatory forgetting, which, in the interest of future harmony, helps cultures and nations surmount negative events in their past. However, as soon as it is a matter not of reconciliation between two ethnic or national identities but of the construction of a common identity, the question arises as to whether the "higher" feeling of belonging does not have to consign *all* the boundaries that arise out of cultural differences to forgetting. Or, to pose the question a bit differently: To what extent does the alternative to the different cultural identities and collective recollections lie in "cosmopolitical forgetting" rather than "global memory"? Can the "universal we" that is constituted in the human rights debate via the memory of horrific deeds (with the Holocaust, Hiroshima, and Rwanda serving as symbols of atrocities committed against humanity as such and thus relevant for each one of us) emerge from a void?[69]

Communication of lived experiences on mobile media and social networks was described in the first chapter as a kind of community that individuals constitute through their "techno-communicative activity with each other." What is noteworthy about this conclusion is the understanding of this communication as an "advanced form of 'dis-membering.'"[70] "Dis-membering" refers to overwriting or "bleaching out" the individual past through the application of a universal written history, as was customary within the Cistercian order, where adult novices' secular biographies were so imbricated with the biblical text universe that every memory of the former was automatically associated with passages from the latter, and an image drawn from court circles, for example, now also called up its spiritual counterpart. This "spiritualization of personal memory images" signifies a "conversion of private memories into memories of the community," that is, not a loss of the memory but its bleaching out under the aegis of a new collective feeling.[71] If we apply this concept in the context of social media, we should therefore ask about the "bleaching" and "spiritualization" that convert the personal into the communal. How might we think forgetting, or alternatively "bleaching," in the context of a digital nation?

The answer takes us on a detour via the community model of Jean-Luc Nancy, who was already mentioned briefly as the author of a philosophical rationale for social networks in the sense of phatic communication and whom we will now consider in greater detail as a possible precursor to thinking about an identityless—and memory-free—community. Nancy's philosophy is shaped by the insight that politically and culturally defined communities are characterized, in the best case, by individual heteronomy and boundedness toward external forces and that, in the worst case, they end in political terror, social violence, and nationalist aggression. His concept of community therefore inquires into the possibility of human beings who would transcend the perspectives and standpoints that distort humanity and aims at a concept of human beings that would be prior to culture: "Can we think an earth and a human such that they would be only what they are—nothing *but* earth and human— and such that they would be none of the various horizons often harbored under these names, none of the 'perspectives' or 'views' *in view* of which we have disfigured humans [*les hommes*] and driven them to despair?"[72] Fundamentally, this is about human beings *as such*, the way the Enlightenment imagined them. Thus, for Nancy as well, the essence of community lies not in a common substance, as something that would be shared by everyone, as communion, but instead in the etymological obligation of the *cum* as being-with-others, a being together that does not presuppose a common being. "What this community has 'lost'—the immanence and the intimacy of a communion— is lost only in the sense that such a 'loss' is constitutive of 'community' itself." In other words, what is felt as a lack proves to be a gain, on which it is necessary to build.[73]

If community is thought prior to any definition of its content, hard and fast referents, or essential characteristics, then phatic communication ultimately also plays an important role. It sustains the conversation, the contact, the being-*with*, precisely by demanding nothing more than the gesture of conversation. Even "chatter" is evidence of the wish "to maintain oneself as 'with' and, as a consequence, to maintain something which, in itself, is not a stable and permanent substance, but rather a sharing and a crossing through."[74] Here, in examining the theme of sharing and connection on an abstract level, Nancy comes as close to Facebook as it was possible to do a decade before its creation. Since then, social networks have variously been described as "phatic technologies" that are more about community building than exchange of information, and Facebook itself has been described as a place where, based on "disinterested interest" in their respective status updates, people develop "disinterested sympathy" for one another. It is even claimed that the resulting "pan-sympathy" surpasses the "natural sympathy" described by David Hume.[75] Is Facebook, then, where connections are created from connections rather than from common interests or values, the site where Nancy's society becomes reality, without anything to bind it together? Is there more behind Zuckerberg's talk of "our community" than self-advertising and self-deception? Is Facebook's "community" the "digital nation" evoked above?

In a certain sense, the "ecstasy of communication," in which, for Nancy, "singular beings" face one another as separate entities without blending together, occurs all the time on Facebook, in the intoxication of status updates. These are connections without commitment, a community lacking ground and work.[76] That longer, more complicated posts receive few likes is perfectly plausible from this perspective, since the person who is seeking content undermines the model of phatic community. The breach lies in the search for a ground of communication that would go beyond being together in the act of sharing. In this regard, the concept of episodic identity, which was introduced and critiqued as an impoverishment of experience in chapter 1 and quasi rehabilitated in its opposition to the narrative type of identity in chapter 2, is ultimately also revealed in its social function. For the narrative individual is a "work" that, secure and self-aware, steps forward to face or confront other individuals: "Individuation detaches closed off entities from a formless ground." To this *individual* immanence, Nancy opposes the "singularity" that emerges out of nothing and returns to nothingness: "It is not a work resulting from an operation. . . . Its birth does not take place *from out of* or *as an effect of.*" Its ground is a "groundless 'ground' . . . [in] that it is made up only of the network, the interweaving, and the sharing of the singularities." From this distinction, the title of Nancy's book *Being Singular Plural* derives its sense of opposition to any culture-pessimistic interpretation à la Sherry Turkle's *Alone Together* (2011). Nancy's ungrounded community of being *with*—herein lies the source of its "ontological 'sociality,'" beyond the "idea of a social being of humankind" as a "zoon politikon"—is a "community of singularities" that, unlike Turkle's "addition of individuals," is no longer conceived sociologically or psychologically but fundamentally already mathematically.[77]

As abstract and idealistic as Nancy's model of community may appear, it represents an important objection of contemporary philosophy against the metaphysical ground of previous concepts of community. At the same time, Nancy's concept appears quite compatible with the existence of the uprooted subject in globalized late capitalism and with the "man without substance" or belonging, like Tikkun's description of the Bloom type: "man who became truly abstract," liberated by social alienation and loneliness to participate in "veritable community."[78] Is Nancy's concept compatible with the globalizing media of late capitalism, as well?

Nancy never connected the question of community with the question of the media and new technologies as clearly as one would have wished. But when he announces the condition of "struction" (which, as "the uncoordinated simultaneity of things, or beings" and "the pure and simple juxtaposition that does not make sense" is equally far removed from construction and deconstruction) as "the lesson of technology," this can be read as a statement about "inserting ourselves into a technosphere," with technology as the central actor in the meaning of history. Then the withdrawal of meaning that Nancy describes appears as a "technological shift of meaning" to the externality of technique, so that it becomes possible to connect it with cybernetic theories about technology as a constitutive environmental factor in human culture. However premature this connection may be, its appeal for thinking about a utopian society beyond cultural differences lies in the prospect that the sense of being is determined technically rather than culturally and that technology itself is not cultural—or, if it is, then only in a global, unifying manner that transcends specific cultures with regionally, nationally, religiously, etc. based differences. In other words, if the *being with* of the inoperative community, if the uncoordinated copresence of *struction* is to be thought, after all, within the framework of a binding context beyond the merely biological, then the point of reference for it would be not culture but technology. The hope behind theories like this is concentrated on the technical unconscious, which, in the information-rich environment of ubiquitous computing and the internet of things, creates a "cybernetic subjectivity" antecedent to the mental and collective structuring power of linguistic signification.[79]

Such a vision of the future can currently only be asserted and sustained at a highly abstract level, which does not have to respond to small-minded calls for evidence drawn from real life. If, in the meantime, we want to seek signs of the overwriting of various cultures by a global technology in the arena of practical life, our attention, once again, will turn first to social networks. Are they the space in which we—as members of a virtual community without any binding foundation—experience ourselves more as human beings than as citizens, more as a singularity than as individuals? Is Facebook the place where it is precisely the postings drawn from concrete life that point toward the abstract humanness of the communication partners? Does politics, as critics of the cybernetic paradigm declare (and complain about), take a back seat to a "being with" and "being next to" that has no meaning?

The assumption that Facebook could be the practical equivalent of Nancy's philosophy of the postpolitical (or even its cybernetic radicalization) initially seems to be contradicted by the impression that in the twenty-first century Facebook is the central site of the political—an impression generated by headlines like "Facebook revolution" in the context of the Arab Spring in 2010–2011, by the manifold possibilities for messaging and mobilization on Facebook, the large number of political Facebook groups, and the fact that in millions of cases profile pictures on Facebook are used to make political statements, in the form of "JE SUIS CHARLIE," tricolor national flags, or—launched in the form of an app by Facebook itself—the rainbow colors of the LGBT community. Naturally, no one will regard Facebook, solely on this account, as the wellspring of a political consciousness that would question the social status quo and look for social and economic alternatives in the sense of critical theory. It is telling that Zuckerberg never portrays Facebook as a tool for political change but always invokes its mission of a general "brotherhood" that transcends concrete political action. And it is unmistakable that when Zuckerberg talks about a universal human right to the internet, he is operating not so much politically as with an eye to the profitability of his company. Facebook's stance may be "libertarian" and "tolerant" in the same way the "postpolitical young people" of the "digital nation" were described some twenty years ago, but when it comes to social philosophy it is definitely conservative.[80]

Facebook assuredly does not question the existing political-economic system. Rather, it secures it in four ways: (1) data analysis makes it possible to personalize advertising, which increases its efficiency by heightening market acceptance; (2) the mix of personal information and advertising accustoms us to seeing ourselves increasingly as part of consumer culture; (3) the possibility of comprehensive control of individual and collective behavior on social networks leads to subtle forms of self-censorship; and (4) the primarily phatic communication and growing extent of nonreflective relations with self and society undermine the intellectual basis for political opposition. Facebook does not operate directly to suppress politics, but it significantly encourages a nonpolitical stance. How detrimental this can be for a culture of political discussion is demonstrated by empirical studies showing that political postings are often ignored or blocked. Even on political Facebook pages, the interest in garnering greater numbers of visitors often means that substantive discussion is supplanted by uncritical agreement with seemingly obvious positions or by sensationalism and simplification. Critical network theory therefore describes the dialectical character of the internet as a space of alternative forms of communication and generation of critical views that, at the same time, is controlled by big corporations and that, as fragmentation of the public realm, represents a new form of "repressive tolerance." To the extent that critical network theory continues to support the notion of communicative deception (*Verblendung*), it also holds fast to the hope that alternative digital technologies, far from commodifying freedom, could have a politically emancipatory effect. On the other hand, if we regard the problem from a more anthropological than political-economic perspective, we are led to suppose that the reason why Facebook is so successful is precisely its effectiveness when it comes to suppressing the real political controversy.[81]

This conclusion does not apply only to Facebook. It is equally valid for other social networks that have a lasting impact on Facebook society. Hossein Derakhshan, who spent six years in prison in Tehran for his political blog, provides a contemporary account of the political impact of these networks. His view of the internet, in the year 2015, is full of disappointment over the "loss of intellectual power and variety" and the untapped potential that the internet "could have for our plague-ridden times." Derakhshan sees the problem specifically in Facebook, Twitter, Instagram, and Snapchat, which are driving the shift from meaningfulness to popularity, from complexity to short-windedness, and from text to image. On social networks, he argues, text is increasingly displaced by videos and pictures, and the decline of reading in favor of seeing and hearing ultimately represses listening as well: "I miss the time when people took the time to be confronted with differing opinions, and were ready to read more than 140 characters."[82]

The fact that Facebook offers a platform for political contents along with everyday banalities does not disguise the fact that its technical and social *dispositif*, by encouraging less reflective forms of communication and a focus on everyday contents, stands opposed—in principle—to a culture of political discussion. In light of Nancy's postpolitical model of community, this depoliticization should not be too quickly dismissed, however. Instead, we should think of it, initially, as

Facebook's actual political function—as a defense, in Nancy's terms, against identification as foundational event. The question of "dis-membering" and "bleaching out," which was posed in the context of social networks, should be answered in precisely this sense (of Nancy's opposition to all forms of collective identity). There is no "spiritualization of personal memory images" taking place in the interest of creating a new relational framework; private memory is not being converted into communal memory. The accompanying lack of political, national, or cultural confession is a confession of a very particular kind, one as constitutive for Facebook society as, according to Nancy, the lack of individual immanence is for society as a whole. The value ascribed to this lack leads to the more profound problem of "identityless identities." In the context of this discussion, two steps are required for this question. Thematically and historically, they take us back to the beginning of this chapter.

In his 1994 essay "Nihilism on the Information Highway: Anonymity vs. Commitment in the Present Age," the American philosopher Hubert L. Dreyfus described the internet as a place without values and meaningful communication. Dreyfus's critique opens by referring to a text that recalls Schopenhauer's remark about the "Babylonian confusion of tongues" among people who read too much: Søren Kierkegaard's *Two Ages: A Literary Review*. In that book, which appeared in 1846, Kierkegaard mounted a critique of the disorientation and paralysis that he claimed afflicted individuals as a result of the overabundance of contradictory opinions and the effect of newspapers and cafés on the expanding public realm. As Dreyfus summarized, "Everything is equal in that nothing matters enough that one would be willing to die for it." Like the daily press for Kierkegaard, for Dreyfus the internet, with its openness to multiple opinions and lack of accountability, is "the ultimate enemy of unconditional commitment." His final conclusion mobilizes three of the most important, sweeping judgments on the internet, calling it "unreal, lonely, and meaningless."[83]

Ten years later, Evgeny Morozov reopened Dreyfus's discussion with his book *The Net Delusion*, in a chapter called "Why Kierkegaard Hates Slacktivism." For Morozov, "slacktivism" —a net-cultural neologism formed from "slacker" and "activism" and carrying the same negative charge as armchair activism—symbolizes the form that modern political "engagement" has taken. A person retweets the link to an anticorporate video because she also hates big business; he posts the yellow umbrella as his profile picture to announce his sympathy with the Hong Kong umbrella movement; she joins an interest group for the defense of the environment because it is chic to belong to it; he signs an online petition or blocks the link to a video by Anonymous. Morozov's critique has been challenged—even Facebook groups that a person joins for tactical reasons can awaken or sharpen political consciousness. But his complaint seems justified at least when "click activism"—to borrow another neologism—is used to justify a person's inaction in reality.[84]

The reference to Kierkegaard, however, is problematic. Kierkegaard's critique of the pluralism of opinions and his praise of unconditional engagement no longer seem appropriate in the era of ideology and language critiques. Would it really be desirable if a combination of distraction and consumerism were no longer able to mask the absence of meaning and if netsurfers were to seek refuge from the despair of meaninglessness in offers of supreme meanings and immovable ideologies—which, after all, are not offered only by Islamic fundamentalists? Or in new legitimating metanarratives that are again worth living for or, if necessary (as Kierkegaard says), dying for?[85] Is it appropriate to accuse the current generation of students of "fuzzy-headed nonsense about tolerance" and "inability to take a position" and to demand that instead of an "undecided-optional 'on the one hand, on the other hand'" they adopt a clear "whoever is not with me is against me"?[86] Can the identity crisis of the postmodern individual still (or once again) be solved using this sort of vocabulary? Naturally, what matters is why and under what circumstances a particular stance is called for. Who could take umbrage at an "unconditional commitment" to oppose poverty, disease, or environmental pollution? The unconditional stance becomes problematic, however, in the context of ideological belief and political consequences, when, as Adorno warns in his essay on commitment, the feeling of being on the right side too easily seems to justify injustice toward "others."[87]

It is this danger of conviction that makes Nancy seek the foundation of a new community outside of a communal common ground, beyond shared (politically, religiously, or culturally determined) convictions, which always have the potential to divide the community. Only the void represented by the absence of such perspectives promises universal accessibility, for the absence of anything essential is also the absence of a boundary separating different essential forms of being. This, admittedly, is precisely the point where critics start to worry about Nancy's "philosophism," which seeks to grasp "being with" *as such*, beyond concrete politics and empirical sociology, and fails to account for the "we" of the hoped-for community in its concrete context. Nancy thinks "being with" primarily from the perspective of the "with," even though the latter is subordinate to being: We are always already in concrete life situations, which tend to work against the "with" not only because in them it is always also a question of access to resources and power but also because we grow up as individuals rather than as singularities. The I's of "being with" are no longer a tabula rasa; the void is lost as soon as it can be named, for with language, the world fills us, and it fills us within the forms and boundaries of this language. Nancy's concept of community lacks a political theory of contention, to clear up all the differences and conflicts that work against formation of the desired community. The critique necessarily also applies to Facebook, which has been viewed here as the practice that corresponds to Nancy's theory. The second part of our evaluation of the lack of political debate on Facebook takes us back to the critique of the narrative paradigm.[88]

The academic discourse on narration as an offer of meaning and of a home even in ethically problematic constellations finds, in light of the decline of narrativity at the end of the twentieth century, that humanity cannot get along without meaning-founding "cultural narratives" but, at the same time, that it should regard them with skepticism, as "cultural constructions." What is called for is a type of narrative that is conscious of its ethical ambivalence, of having to serve as both a source of orientation and a blockade that keeps out alternative perspectives: "metanarrative reflections on the tension between the infinite complexity of experience and our necessarily selective narrative accounts." What is called for are dialogical and self-critical "metanarratives," under the sign of pluralistic interpretation. The goal is not the end of stories but the end of their innocence.[89]

This demand asks more than it admits, for it fails to discuss the extent to which narratives, if they are self-critical, are actually able to create meaning and identity. Is it still possible to believe, after Nathan's critique of origins and his self-skepticism, that a person can feel "grounded" in this way? Won't a person be more likely to hold on, like Dajah, to life in the sole possible truth? The demand for self-reflective metanarration leads logically to Vattimo's concept of the "'weakened' experience of truth" and a hermeneutic model that includes an understanding from plural perspectives—an understanding like the one Vattimo develops, in the context of the postmodern critique of reason, with his concepts of "weak thinking" and the "nihilistic vocation of hermeneutics."[90] This thinking is "weak" because it also reflects on its cultural and social preconditions, which constitute humans *as such*, and because, based on this insight, it retains a solid skepticism toward itself. The proposed hermeneutics is "nihilistic" because it will accept no interpretation as *the true* one, from which any deviation would simply be an error. On this basis, every statement, every adherence to a position or profession of belief occurs under the sign of irony and relativity, with an ethical, highly political purpose: "Thinking that no longer understands itself as the recognition and acceptance of an objective authoritarian foundation will develop a new sense of responsibility as ready and able, literally, to respond to others whom, insofar as it is not founded on the sternal structure of Being, it knows to be its 'provenance.'"⁹¹

Does Facebook, as a social network and as a company, practice this type of response, which would exemplify a new sense of responsibility? Does spontaneous, narration-free communication on Facebook provide a corrective to the strong, steadfast thinking promoted by Kierkegaard, Schopenhauer, Fichte, and most likely also by contemporary critics of "slacktivism" and "fuzzy-headed nonsense about tolerance"? Does sharing, as the central rule of Facebook culture, founded on nothing but the desire to connect, encourage a feeling of "being-with" regardless of whether the participants share a common substance? How ethical can phatic communication be?

The problem is antecedent to the act of answering: Phatic communication has no time to listen. Its nature is to be ultimately disinterested in the person it confronts. Communication as "being-with" ultimately lacks "being with for Others," as Martin Heidegger once defined hearing: as the "primary and authentic way in which Dasein is open for its ownmost potentiality-for-Being." The pressure to "share" on Facebook—and this is true increasingly of Facebook society in general—values speaking over listening in a profoundly impatient, not infrequently self-infatuated way that usually prefers snippy comments to serious reactions. The mocking, ironic stance underlying phatic communication sabotages the philosophically grounded irony that results from the acknowledgment of alternate perspectives. Empathy, understanding, comparison—these are the tools of "weak thinking," whose strength lies precisely in the turn toward the other. Vattimo speaks of the "liberation of local rationalities" as the insight that "in a world of dialects" my dialect "is not the only 'language,' but that it is precisely one amongst many." The irony that springs from this acknowledgment knows whereof it speaks. This irony is not playful, ignorant, or mocking but serious and enlightened. It confronts the other not with indifference but with awareness of equal validity.[92]

When Mark Zuckerberg tirelessly refers to Facebook as "our community" and declares that its mission is to increase worldwide understanding, he is by no means talking about taking an ironic approach to narration and identity. Facebook pursues its mission of global embrace by screening out everything that separates people—above all political, ideological, or religious convictions and comments. Communication on Facebook operates in the phatic mode; it flows past as the kind of pleasant, information-free white noise that is where we have surmised the *factual* cosmopolitanism of Facebook society is to be found. To this we should add that while exchanging phatic communications may not mean that we are "learning" a system, it also does not mean that we are learning to disbelieve in any system. The next question is whether cosmopolitanism, as a conscious position that could be sustained even in changing social constellations, can be the "side effect" of an objective, more or less unconsciously or conceivably even intentionally imposed process of cosmopolitanization and thus enter through the "back door," as it were.[93]

Without empirical data, skepticism is the recommended response. That factual "cosmopolitanism" on social networks does not promise or presage any long-lasting cosmopolitanization can be seen when the exchange of banalities is disrupted, now and again, by conflict between different political or religions viewpoints, and when the busy quiet of general indifference gives way to a usually completely unironic, often self-righteous, not infrequently aggressive style of discussion that occasionally escalates into a virtual lynching. Web 2.0 does not seem to cultivate acceptance of the Other in any confrontation, something that is hardly astonishing, given a culture of sharing and delegated enjoyment in which the model of reflective experience over time is replaced by a phatic model of short-term lived experience. Distanced, differentiated, self-critical discussion is the first victim of this shift. The question is not whether Zuckerberg the businessman believes in Facebook's mission but whether Facebook's technical and social *dispositif* is constructive. The doubts follow from the opportunism of phatic communication, which achieves the utopia of generalized understanding only by excluding everything that is in danger of drawing boundaries. The phatic element does not offer protection against the outbreak of new "truths" whenever the flight into hyperactive distraction—and happiness in the mode of consumer culture—no longer succeeds.

Basically, Facebook, with its affirmative like(able) culture, is a big feel-good party that can be compared to certain participatory art projects that, since the end of the twentieth century, have aimed to produce, on an aesthetic level, something social networks have been practicing on the cultural level since the beginning of the twenty-first. The critique that has been leveled at this form of "relational aesthetics" in the debate over art should be addressed to Facebook culture as well: The foundations of a democratic society are secured by reflection and cognition, not sensation and immersion. Even a deconstructive subjectivity must be construed with awareness of its context. How this happens concretely is a question that should be asked of Nancy's concept of community, as well. A start, for social networks, could well be to stop misunderstanding the bond that is sought and experienced in these exchanges as the shared celebration of individual existence and instead see it as a reaction to a shared deficit, in each case, when it comes to the meaning—Nancy would say the "grounding"—of life and as an insistence on holding fast to this deficit as its actual sense.[94]

AFTERWORD

"**N**ever before has an age been so informed about itself, if being informed means having an image of objects that resembles them in a photographic sense." Kracauer immediately contradicts this praise of photography when, a few lines later, he characterizes it as a "strike against cognition," because reproducing reality mechanically makes it superfluous to grasp it consciously. Kracauer's commentary on photography provoked a discussion of the cognitive achievement of social networks and the informativeness of Facebook society. What we found was that the photographic form of self-representation delegates individual experiences to the social network and suppresses narrative forms of perception. This conclusion may seem less threatening in light of the critique levied against the narrative mode and given the corresponding advantages of forgetting. In the context of this critique, the episodic, phatic model of communication on Facebook was advanced as practicing a model of community that transcends divisive narratives and identity constructions. This unorthodox perspective, then, took yet another turn with the call for "weak thinking," as habitual tolerance that results from "working through" conflict, as opposed to unreliable indifference, avoiding all conflicts, in the model of phatic communication. "Weak thinking" responds to discredited narratives not with opposing stories but as a "story" that is opposed to stories. It is the alternative, less popular reaction to the vacuum left by the loss of traditionally meaning-creating stories: an alternative to self-satisfied communication on Facebook, which enjoys the popularity it does because it is experienced as liberation from communication that is weighed down by meaning. This liberation is the site of the "go-for-broke" gamble of history, which Kracauer, ninety years ago, wrote about in relation to photography.

From his description of the increasing popularity of photography as a loss of society's knowledge of itself, Kracauer derived a surprising prognosis. Photography, he said, makes society fall silent because in it the material itself, bypassing its meanings, speaks as a "barren self-presentation of spatial and temporal elements." In doing so, photography, he claimed, frees consciousness from the narrative orders given to things by human beings and brings it into direct contact with nature. This liberation makes it possible to reframe meaning, in order to "awaken an inkling of the right order of the inventory of nature," presuming that "a society that has succumbed to mute nature" does not persist. The risk, then, lies in muteness persisting after all, which, for Kracauer, would mean the "eradication" of consciousness: "The turn to photography is the *go-for-broke gamble* of history."[1]

A medium as a game of chance? Is this metaphor anything more than slyly formulated cultural pessimism? Does it anticipate Walter Benjamin's "positive concept of barbarism," with which, five years later, the latter would greet the gambling away of the "human heritage" for the "small change of the 'contemporary'" as "making a new start"?[2] Is history, today, once again betting everything on a single card—with objective forms of self- and world representation that elevate the "foundation of nature devoid of meaning" characteristic of photographic documentation—going all in on the operating mode of Facebook society?

The barbarism of the new lies in gambling away narrative, reflective consciousness, which is increasingly being suppressed by numerical, visual, and automatized forms of communication and types of information processing. Central factors in this development are the methods of quantification found in self-tracking and Big Data mining, along with visualization technologies like Snapchat and future VR/AR technologies like Facebook's Oculus. The future is "frictionless sharing," without reflection and ultimately also without control by the sharers. This is precisely what, to some, promises the end of the distortion produced by the subjective bias of narrators and the compulsory coherence imposed by the narratives. For these observers, the outcome promises access to a "mystery of being" that transcends narrative concepts of explanation; for them, the episodic model of identity is welcome as the end of exclusion and heteronomy through collective narration.

The exclusion of the I from first-person narrative is the paradoxical equivalent of the self-presentation of the material in photography. With the new technologies, the self-presentation of the I occurs without its conscious participation. This posthuman self-abnegation of humanity is comparable to the nature, that, as Kracauer has it, sits down at the table consciousness has just vacated. If it were to become the new head of household, in analogy to Kracauer, history would have lost everything. For the outsourcing of narrative to alien authorities also means the abandonment of the practice of reflection—a loss that should not be seen, somehow, as a technical translation of the complex concept of "weak thinking," which can be difficult to convey, but that instead destroys any foundation for it. Could our hope lie in the return of the old, to resume its seat at the table alongside the new?

The linking of the numerical and the narrative, of algorithmic analysis and hermeneutical techniques, is *the* contemporary topic in the realm that falls most essentially to hermeneutics and narration: the humanities. The catchword for a humanities that would be dedicated to algorithmic methods of analysis is "digital humanities"; the anxiety-inducing words are "distant reading" and "the end of theory." A hint of reconciliation is bruited about in the concepts "algorithmic criticism" and "ecology of collaborating."[3] There are, somewhat simplified, three camps: those who hold fast to the process of interpretation as the most integral and essential method of the humanities, those who want to produce authoritative knowledge by means of quantifying data analysis, and those who expect data mining to produce new approaches to the business of interpretation. The third camp also sees the new means of data collection as a source of challenges and opportunities for our understanding of rationality, consciousness, and self-experience. This group remains committed to narration but does not exclude counting from recounting. It allows itself to be inspired, but not corrupted, by the new. Finally, it defends the position of "weak thinking" and "nihilist hermeneutics" against the "spatial appearance" and "barren self-promotion of 'facts'" and tries, when it comes to social networks, to combine the insights of posthuman self-representation with aspects of the cognitive activity of conscious self-description.

The digitalization of the humanities is itself part of a "transformation of the human" that adds a wholly new dimension to such general humanities themes as reason, consciousness, and self-understanding. Two of this transformation's catchwords were already mentioned in the context of Facebook's posthuman, algorithmic autobiography: "nonconscious cognition" and "distributed cognition environments."[4] The outsourcing of narration and recollection and the shift from narrative to numbers on social networks and in the humanities are phenomena and building blocks of a development that originated long before Facebook and that point far beyond the concept of Facebook society. Material things, which today—in both nature and society—are beginning to present themselves in a way that bypasses people, no longer consist only of the objects in a photograph but also of the data in the feedback loops of cybernetics. The "barren self-promotion of the spatial and temporal elements" that Kracauer saw in photography is now happening in the logic and from the perspective of cybernetics, which is a logic of computation and decision making, of analysis and control, of conditionality and lack of discussion—it is the paradigm of logocentrism in the form of the numerical. The stakes being gambled with in the twenty-first century, fundamentally, are mathematics.

This new risk has a long prehistory. Half a century ago, framed as "technocratic rationality," it was a popular target of critique in the humanities and already formed the basic theme of *Dialectic of Enlightenment* (1944), as Theodor W. Adorno and Max Horkheimer discussed the transformation of rationality from a means of human emancipation to a means of its reification. If we translate this sociological observation from the past into the technological determination of the present, which gushes enthusiastically about cybernetic recursivity and "deep learning" algorithms, reification amounts to the control of humankind by the artificial intelligence that humans themselves have created. This disempowerment was already illustrated by Stanley Kubrick's 1968 film *2001: A Space Odyssey*, in which a computer locks man out in space, and more recently in Alex Garland's film *Ex Machina* (2015), when the robot locks the human being in a room. The new Turing test consists not in the computer convincing us that it is a person but in its convincing us that as a machine it is nevertheless acting like a human. If we believe it—at least this is the upshot in *Ex Machina*—we have/ are lost.

Benjamin's one-time praise of barbarians is expressed, in the current constellation, as enthusiasm for the "technical intensification of complexity" or "disenchantment of the

Anthropocene's control fantasies" and as critique of any "negativistic media theory" that complains about the "cybernetization of the means of existence" as being equivalent to human heteronomy and the mathematical reduction of humanity. What is astonishing about this position, which does not lose any time worrying about the shift in control *among* humans (catchwords: cybercrime, hacked control systems), is not so much the joy elicited by this "fourth insult to humanity, following Copernicus, Darwin, and Freud" as its timing. Humanity's exceptional position is called into question at the moment of its greatest triumph, when it has advanced the capacity for thought given to it by nature to such a degree that it can now pass this capacity on. This passing on, the delegation of the tasks of control to the environment and the application of artificial intelligence, can only be understood as a *loss* of power and an insult to humanity if one suspects that operational accidents—the shutting out or locking up of humanity—are the rule, a position that reveals the individual who thinks this way to be either a cultural pessimist or, if accompanying feelings of joy are to be taken seriously, a cynic.[5]

The promise inherent in the gamble becomes clearer if "environmental cybernetics" is seen as an "epistemological and ontological correction" not of humanity's predominance but of the human dilemma: the dilemma of being entangled in narratives, in "perspectives" or "views" "*in view* of which we have disfigured humans [*les hommes*] and driven them to despair," in Nancy's formulation. Can cybernetics, if it actually replaces politics with technology and does not just dress politics up technologically, correct for this entanglement, since its mode of operation, which does not recount but only counts, can't be influenced by perspectives and views? Can we imagine that in the so-called state of nature of cybernetic control circuits, autonomous artificial intelligences will neither include nor exclude humans but merely act as partners and educators that help humans to be "nothing *but* earth and human," to borrow Nancy's description of the alternative? If we, then, start by thinking the technological determinant together with Nancy's concept of community, does the historical point of cybernetic transhumanism consist in humanizing the human by technological means?[6]

The twofold outcome of the gamble is the topic of differing philosophies of technology. Heidegger's enframing (*Ge-stell*) is a scaffolding that offers support but also limits movement; Stiegler's *Pharmakon*, depending on how it is used and on its dosage, is either poison or medicine; even cybernetics has a dual value, as "left" or "right" cybernetics, depending on whether it is viewed as static and system preserving or as creative, learning-friendly autopoiesis.[7] The future will show how the new go-for-broke gamble of history can be won and how the "cybernetic state of nature" that is evoked (its structure, mode of operation, information sources, knowledge criteria, levels of complexity, and rules of recursion) can be thought concretely. In any case, as this book has discussed, a life that is surrounded and besieged by numbers (as we might tendentiously describe "ubiquitous computing") is already the permanent object of archiving and surveying. Already today, algorithms are filtering a closed system of knowledge from out of life. Perhaps at some point, using AR/VR technologies, they will be able to avoid culturally determined conflicts by cleverly interposing individual parallel worlds. Artificial intelligence, in popular forms like Siri, Alexa, or Jibo, is already part of our activities and will only become more and more active thanks to the input of social media.

If we turn from speculation about future technological constellations to an analysis of contemporary media interactions, the result of our discussion of Facebook and Facebook society can be summarized as follows:

First, permanent talk about oneself on social media is flight from the events occurring in a person's life; we are exhibitionistic not because we are narcissists but because we cannot bear ourselves and the present. Sharing on Facebook should be understood as a stopgap; it gives us a decent option for delegating our own experiences to others. Second, self-representation on Facebook happens less in a way that is narratively reflective than as a spontaneously episodic and documentary event. The outcome is a quasi-automatic autobiography whose central narrative authority is the network with its algorithms. The self-image that is presented by Facebook is pointillist, postmodern, and posthuman. Finally, information management on Facebook and on the internet suppresses collective memory. With its lack of narrative points of reference in the framework of phatic communication, it creates a quasi-cosmopolitan community that transcends cultural values and national barriers. However, at the same time, the avoidance of discursive interaction prevents the development of skeptical, metareflexive thought as long-term security against new forms of assertive dogmatism.

Let me add one more thing. Even a description of society from the perspective of cultural studies can hardly avoid being drawn in by the perspective of its speakers and the force of coherence that their narrative exerts. No analysis of the present can escape the analyst's past. Thus, in the end, I may appear more critical than I wished to be when I began. But at the latest, when ten thousand kilometers away my students enthusiastically report (grinning, it is true, but ultimately without any awareness of guilt) how much time they have spent on Weibo with their "idols and celebrities" and in sharing bits of knowledge like "How you can freeze a can of Coke very fast," I see clearly that outside my own culture of thought there is an entirely different relationship to the world and its media—an agreement in principle, a childlike enthusiasm even, unburdened by social-political ambitions and left melancholy. Then, at the latest, I cannot avoid the sense that everything can be viewed entirely differently. Then, at the latest, it is high time to give the word to the voice that opened this analysis:

The food arrived and, as always, this was the most exciting moment. Everyone grabbed their camera; together they rearranged the plates for the perfect photo. "Seung gei sik see" is the saying in Hong Kong: The camera eats first. The camera is the modern saying of grace, the sharing of "bread" in its symbolic representation. For now, the first thing that happens is the passing around of the result, the exchanging of the best images, their sharing on Weibo, Facebook, Instagram . . . Then the eating can begin. The last morsels haven't been eaten, and already the first results are in. This, too, is cause for all kinds of conversation. Everyone knows the same friends, who now send their "foodies" from the places where they are eating, alone or together. A complex dialogue emerges among the images and texts, full of revelations and inside jokes. The foodie is not proof of loneliness in a crowd, as cultural pessimists like to think, but a vehicle for communicative action that is full of fun and deeper meaning among different groups, in different places, via mobile media and social networks. This communication accompanies every bite and later continues for days, in commentaries on Facebook or Weibo. It is engaged and generous, for it includes pizza as much as oysters. The foodie demonstrates food not as an object but as an action. It creates community through its link to the most essential things, to that which joins all human beings without regard for their political position and cultural values. What the mass media manage to do with murder cases and reports of catastrophes, the social networks achieve with everyday banalities. The only blood that flows here comes from the steak.

EPILOGUE TO THE
ENGLISH EDITION

A book about a fast-moving digital technology risks being out of date almost before it appears. But if the book is a philosophical treatment rather than a manual or an empirical study, it is less likely to suffer from premature obsolescence. For this kind of study, new developments are welcome, for they offer a way to gauge whether the book's philosophical arguments and speculations about the deeper meaning of a given technology are actually confirmed.

A main thesis of this book is that social networks and diary apps prompt their users to engage in more or less unconscious and unreflective self-narration of a kind that favors implicit over explicit self-revelation and that prefers mechanical presentation (via photography or automated sharing) to mindful representation (via textual statements or the creation of a narrative structure). The mode of self-expression that takes place on Facebook and other social networks is spontanous, episodic, and documental, rather than deliberate, coherent, and narrative. As a result, it generates a kind of "automatic autobiography" or "posthuman self-description" whose actual narrators are the network and its algorithms.

Evidence for this argument has continued to accumulate since the book appeared. Just a few weeks before the release of the original German edition, Mark Zuckerberg elaborated on his vision of the "next big thing," the video-streaming feature Facebook Live, which launched in April 2016. "Because it's live, there is no way it can be curated," he said. "And because of that it frees people up to be themselves. It's live; it can't possibly be perfectly planned out ahead of time. Somewhat counterintuitively, it's a great medium for sharing raw and visceral content."[1]

The buzzwords for this new form of frictionless sharing are *raw content* and, implicitly, *transparency* and *truth*. Like self-tracking and "numerical narrative" (discussed in chapter 2), "frictionless sharing" seems to offer users a "cure" for the inclination to narrate their life in a way that makes sense to them and that inevitably changes it in the process. The desire for more "authentic" data from Facebook users also targets the *viewers* of a Facebook Live video, who, as with Twitter's live-streaming video app Periscope, launched in March 2015, can instantly and spontaneously add a thumbs-up or a comment. Judgment is rendered while the video is still streaming; before Periscope, the viewers would have had to wait until the entire video had been recorded and uploaded, then downloaded and watched before they could add their comments. Spontaneity and immediacy function here as synonyms for authenticity; they are the counterpart to frictionless sharing.

An app that already, and very successfully, allows uncensored acting in public is Snapchat, which famously promises that the images shared on it will self-delete after they are viewed. I identify Snapchat as the logical next step after Facebook, since Snapchat not only destroys the experience of the present by constantly capturing and sharing it but also abandons the archive, which has become dispensable in this dialectic of preservation as forgetting. Since then, Snapchat has gone public, and, in March 2017, Facebook copied Snapchat's signature feature with its Messenger function Day, which deletes images and videos after twenty-four hours. The fact that the lifespan of the image is almost as short as the time it takes to produce it perfectly suits the desire to erase the experienced moment by capturing and sharing it—and never coming back to it again. An idea as seminal as this is too good to be left to the competition.

These new features confirm the prognosis of this book that the type of self-representation to which social networks increasingly seduce us provides less and less content for our own self-reflection and self-understanding and more and more reliable material for the algorithms at the back end of the interface. Not only do we favor the camera as the means of mechanically reproducing our realities; we don't even take the time to manipulate the images we upload. In essence, we increasingly cease to be the authors of our own autobiographies.

Another central concern of this book was to take a close look at the idea of a digital nation or cosmopolitan community as something the internet or Facebook claims to be able to generate. The book points out that Facebook pursues Zuckerberg's declared mission of creating a "global community" by favoring postpolitical phatic communication—a model that, at the end of chapter 3, was discussed as the praxis conforming to Jean-Luc Nancy's theory of a groundless community. We concluded with the notion that today algorithms are already filtering a closed system of knowledge from out of life, and speculated that at some point they might be able to avoid culturally determined conflicts by cleverly interposing individual parallel worlds.

In February 2017, Zuckerberg presented his manifesto "Building Global Community." The statement stressed the political aspect of postpolitical communication and declared that "the best solutions for improving discourse may come from getting to know each other as whole people instead of just opinions."

It will be interesting to see how Zuckerberg responds when opinions—above all political, cultural, or religious convictions—actually do separate people. While, as we argued in the discussion of social media and community, Nancy's concept of "groundless community" lacks a political theory of contention, Zuckerberg's solutions for all the differences and conflicts that work against the formation of the desired community consist in algorithms and filter bubbles.[2]

Holding that for a community of two billion people it is not feasible to have a single set of standards that governs all divergent opinions, Zuckerberg suggests that "we need to evolve towards a system of personal control over our experience." The objective is not a mutual experience of different perspectives but a customization that conforms to a person's own point of view. "Each person should see as little objectionable content as possible," Zuckerberg promises. To him, of course, this form of "self-governance" epitomizes the expansion not of the filter bubble but of "democratic referenda."

It could be tempting, on first reading, to agree with Zuckerberg. By treating humans as biological beings and affording them the possibility to decide freely, for themselves, outside cultural contexts and belief systems, he avoids prescribing universal values for all those who coinhabit his "global community." But the utopia of general understanding cannot be brought about by excluding everything that might result in the drawing of a boundary line. Shutting the Other out of sight and mind is not a form of tolerance. Tolerance, rather, means putting up with difference. To strengthen the unbounded We of humanity, writ large, against the We of nations, cultures, and other forms of belonging, it is not enough to liberate egos from their previous forms of groupthink and to set them loose, through a system of personal control, in their own filter bubbles. Global community must be acquainted with itself in all its facets. In other words, individuals' freedom to decide their own lives must ultimately be restricted when it comes to the lives of others.

What is disturbing about Zuckerberg's manifesto is that he thinks the "engineering mindset" he credits himself as having permits him to solve a problem as complex as universal human rights and global community, and to do so through the quick application of a few technical fixes. But hyperlinks do not necessarily lead to understanding, and transparency does not necessarily end in empathy. The initial premise of the manifesto— "Facebook stands for bringing us closer together and building a global community. When we began, this idea was not controversial"—betrays ignorance of the controversial debates over multiculturalism and cosmpolitanism. To this is added a complete lack of reflection on the extent to which the envisioned technical means are actually constructive when it comes to realizing the goal he envisions. Building a global community is at least as complicated as health care.

Toward the end of 2016, there was increasing speculation about Zuckerberg's ambitions to hold America's highest political office. A President Zuckerberg would be the correction to Trump, for his motto is not "America First" but the whole world. His model is not polarization and re-ideologification but linking and small talk. A President Zuckerberg would certainly be better than a President Trump. But would a Facebook society be good?

In his February 2017 manifesto, Zuckerberg lays out his purpose as fighting "sensationalism and polarization leading to a loss of common understanding." Naturally, he does not concede that it is Facebook's mode of communication that hinders nuanced, well-thought-out conversation: the dualistic reaction scheme of likes and dislikes; the number-based populism; the time pressure under which contributions are received, evaluated, and recommended. If we consider that the medial mode of Facebook communication encourages neither a reflective world image nor a reflective self-image, it must be doubted whether Facebook will bring about the society that Zuckerberg promised in his manifesto. But perhaps it is too early for this kind of judgment. We will have to continue to keep an eye on how things develop.

NOTES

PREFACE

1. In the subway in Hong Kong the following audio loop can be heard on the escalators: "Please hold the hand rail. Don't keep your eyes only on the mobile phone!" The police in Lausanne alarm pedestrians with a video that warns them about the fatal consequences of texting in the middle of traffic: http://www.youtube.com/watch?v=D-FZI1301Ko.
2. Johann Adam Bergk, *Die Kunst, Bücher zu lesen* [The art of reading books] (Jena: Hempel, 1799), 86.
3. An example of the new philosophy of world affirmation is the vitalist, orgiastic scientific and social theory of the French sociologist Michael Maffesoli. Another is the antihermeneutical concept of presence culture put forward by the German cultural studies professor Hans Ulrich Gumbrecht. Both implicitly or explicitly reject the duty imposed by critical theory to improve oneself and the world. It is no surprise that Mark Zuckerberg, the founder of Facebook, also takes a positive view of the situation and, at the same time, praises his own enterprise as a means of further improvement: "While headlines often focus on what's wrong, in many ways the world is getting better. Health is improving. Poverty is shrinking. Knowledge is growing. People are connecting. Technological progress in every field means your life should be dramatically better than ours today." Mark Zuckerberg and Priscilla Chan, "A Letter to Our Daughter," December 1, 2015, http://www.facebook.com/notes/mark-zuckerberg/a-letter-to-our-daughter/10153375081581634.

4. Taylor, "The Politics of Recognition," in *Multiculturalism: Examining the Politics of Recognition*, ed. Charles Taylor and Amy Gutmann (Princeton, NJ: Princeton University Press, 1994), 72–73.

5. "I shall call an apparatus [*dispositif*] literally anything that has in some way the capacity to capture, orient, determine, intercept, model, control or secure the gestures, behaviors, opinions or discourses of living beings." Georgio Agamben, "What Is an Apparatus?" in *"What Is an Apparatus?" and Other Essays*, trans. David Kishik and Stefan Petadella (Stanford, CA: Stanford University Press, 2009), 14. The concept of the *dispositif* (translated in the title of the book as "apparatus") was invented by the French philosopher Michel Foucault as the term for a network of discourses, institutions, laws, practices, and mechanisms for the regulation of phenomena (sexuality, normality, truth, power) and for the formation, administration, and control of subjects. When it comes to social networks, it makes sense to differentiate between technical and social *dispositifs* as they interact with the software or, alternatively, with the network's users. The technical *dispositif* of Facebook includes the quantifiability of reception and interaction in the form of likes and shares, the possibility of lateral connections through links and tags, and the personalized filtering of designated "news." The social *dispositif* includes the imperative to share and connect; the trend toward positive, euphemistic announcements; and the laws of attentional economy, which, for example, lead to likes being given out above all for postings that are easy to understand (visual) rather than complicated or complex (verbal).

6. For phenomena like Facebook or MySpace, Orkut, QQ, and Weibo, the concept "social network" has gained acceptance; occasionally, the definition is given more specificity, as "online social network" or "social network site" (SNS). It might be objected that these are actually frameworks and that the millions of users of such a platform are creating a network only in an emphatically metaphorical sense. It is in this very sense that Mark Zuckerberg talks about Facebook as "our community" instead of "our communities," the same way citizens of a state are imagined, despite the diversity of their concrete groupings, as *one* community (or at least society). In the present analysis, this communal aspect of Facebook is regarded as a technical (and social) bracket for communities or networks at the micro level. Where this essay occasionally uses the term "digital media" instead of social

networks, it is in order to include forms of interaction such as Google, Wikipedia, Skype, Dropbox, etc.

7. Georgio Agamben, "What Is the Contemporary?" in *What Is an Apparatus?" and Other Essays*, 45: "The ones who can call themselves contemporary are only those who do not allow themselves to be blinded by the lights of the century, and so manage to get a glimpse of the shadows in those lights, of their intimate obscurity." It is in this sense, and with reference to Agamben's essay, that Koepenick employs the concept of "unconditional contemporaneity"; see Lutz Koepenick, *On Slowness: Toward an Aesthetic of the Contemporary* (New York: Columbia University Press, 2014), 3.

8. William Davis, "Mark Zuckerberg and the End of Language," *Atlantic*, September 11, 2015, http://www.theatlantic.com/technology/archive /2015/09/silicon-valley-telepathy-wearables/404641: "The boom in affective computing and wearables . . . is driven by the promise of access to 'real' emotions and 'real' desires, accompanied by ways of transmitting these via non-verbal codes." Zuckerberg's prediction can also be found here. On "mathematicised thinking" and the cybernetic paradigm, see Dieter Mersch, *Ordo ab chao—Order from Noise* (Zurich: Diaphanes, 2013), 47.

9. Similarly: "While rational concepts seek unity, 'reductio ad unum' (Comte), intuition, embracing what is multiple, allows us to comprehend the diverse." Michel Maffesoli, "Erotic Knowledge," *Secessio* 1, no. 2 (Fall 2012), http://secessio. com/vol-1-no-2/erotic-knowledge.

1. STRANGER FRIENDS

1. Richard Sennett, *The Fall of Public Man* (New York: Knopf, 1977).
2. See also Sherry Turkle, "Identität in virtueller Realität. Multi User Dungeons als Identity Workshops" [Identity in virtual reality: Multiuser dungeons as identity workshops], in *Kursbuch Internet. Anschlüsse an Wirtschaft und Politik, Wissenschaft und Kultur* [Kursbuch new media: Trends in economy and politics, science and culture], ed. Stefan Bollmann and Christiane Heibach (Mannheim: Bollmann, 1996), 315–31. In 2003, Jenny Sunden invented the phrase "typing oneself into being" for the construction of the subject in virtual space. Jenny Sunden, *Material Virtualities: Approaching Online Textual Embodiment* (New York: Peter Lang, 2003), 3. Theater and the masks that are

associated with it offer obvious metaphors for virtual space, with the masks standing in for "our true self." See Erving Goffman, *The Presentation of Self in Everyday Life* (New York: Doubleday, 1959); Brenda Laurel, *Computers as Theater*, 2nd ed. (Indianapolis, IN: Addison-Wesley Professional, 2013). For a philosophical consideration of the play of identity in Second Life, see Marya Schechtman, "The Story of My (Second) Life: Virtual Worlds and Narrative Identity," *Philosophy & Technology* 25, no. 3 (2012): 329–43.

3. Mark Zuckerberg, speaking at the Y Combinator Startup School in Stanford, CA, on October 20, 2012, http://www.youtube.com/watch ?v=5bJi7k-y1Lo.

4. Danah Boyd, *It's Complicated: The Social Lives of Networked Teens* (New Haven, CT: Yale University Press, 2014), 46.

5. "We were going to live online. It was going to be extraordinary," writes Zadie Smith in her description of David Fincher's film *The Social Network*. "Yet what kind of living is this? Step back from your Facebook Wall for a moment: Doesn't it, suddenly, look a little ridiculous? Your life in this format?" Zadie Smith, "Generation Why," *New York Times Book Review*, November 25, 2010, http://www .nybooks.com/articles/2010/11/25/generation-why.

6. Robin Dunbar, *How Many Friends Does One Person Need?: Dunbar's Number and Other Evolutionary Quirks* (Cambridge, MA: Harvard University Press, 2010); Zuckerberg at the Y Combinator Startup School.

7. Aleida Assmann, "Hier bin ich, wo bist du? Einsamkeit im Kommunikationszeitalter" [Here I am, where are you? Loneliness in the age of communication], *Mittelweg* 36, no. 1 (2011): 15, 22.

8. The reference is to Immanuel Kant's concept of "disinterested pleasure" (*interessenloses Wohlgefallen*) as the criterion of aesthetic pleasure.—Trans.

9. That global linkage is at least part of what social networks claim for themselves is already demonstrated by the name of the first social network, sixdegrees.com, created in 1997. The name is a play on the idea that every human being in the world is connected to every other human being by a chain of acquaintances that have no more than "six degrees of separation." On Facebook's mission of general linking, see also Peace.Facebook.com, which contains information about Facebook friendships between groups that are traditional enemies, and the following comment by Facebook's COO Sheryl Sandberg: "Is it

harder to shoot at someone who you've connected to personally? Yeah. Is it harder to hate when you've seen pictures of that person's kids? We think the answer is yes." Dan Fletcher, "How Facebook Is Redefining Privacy," *Time*, May 20, 2010, http://content.time.com/time/maga zine/article/0,9171,1990798,00.html. Zuckerberg offered a variation on this statement on September 26, 2015, in the context of a UN meeting: "A 'like' or a post won't stop a tank or a bullet, but when people are connected, we have a chance to build a common global community with a shared understanding." http://indianexpress.com/article/technology /tech-news-technology/mark-zuckerberg-calls-for-universal-internet -access-at-un-summit.

10. Friedrich Daniel Ernst Schleiermacher, *Werke, Auswahl in vier Bänden* [Works: Selection in four volumes], ed. Otto Braun and Johannes Bauer (Leipzig: F. Eckdart, 1913), 4. The following quotation is from the same source, 3–4.

11. The speaker is Ulrich, referring to Diotima's salon, in Robert Musil, *The Man Without Qualities*, trans. Sophie Wilkins and Burton Pike (New York: Knopf, 1995), 186.

12. Eli Pariser, *The Filter Bubble: What the Internet Is Hiding from You* (New York: Penguin, 2011), describes information filtering by the algorithms in search engines and social networks such as Facebook as "filter bubbles" and "you-loops" of autopropaganda. Ethan Zuckerman's argument is relatively critical but focuses on the filtering that is done by users themselves. Ethan Zuckerman, *Rewire: Digital Cosmopolitans in the Age of Connection* (New York: Norton, 2013).

13. Seventy-three percent of social network users seldom or never agree with the political postings of their friends. Lee Raine and Aaron Smith, "Social Networking Sites and Politics," *PEW Reports* (March 12, 2012), http://www.pewinternet.org/~/media/Files/Reports/2012 /PIP_SNS_and_politics.pdf.

14. Heinz Buddemeier, "Was wird im CyberSpace aus den sozialen Beziehungen?" [What becomes of social relations in cyberspace?], in *Cyber-Space. Virtual Reality, Fortschritt und Gefahr einer innovativen Technik* [Cyperspace: Virtual reality, progress, and danger of an innovative technology], ed. Horst F. Wedde (Stuttgart: Urachhaus, 1996), 51.

15. Roger Scruton, "Hiding Behind the Screen," *New Atlantis*, Summer 2010. On future relations with robots, see Sherry Turkle: *Alone Together: Why We Expect More from Technology and Less from Each Other* (New York: Basic Books, 2011).

16. Zygmunt Bauman, *Liquid Love: On the Frailty of Human Bonds* (Cambridge, MA: Polity, 2003), xii. Long before Bauman, Lasch made a similar observation: "Our society . . . has made deep and lasting friendships, love affairs, and marriages increasingly difficult to achieve." Christopher Lasch, *The Culture of Narcissism: American Life in the Age of Diminishing Expectations* (New York: Norton, 1979), 30. Latour noted the mutually limiting and reinforcing nature of society and technology: Bruno Latour, "Technology Is Society Made Durable," *Sociological Review* 38, no. 1 (1990): 103–31.

17. Friedrich Schiller, "Die Bürgschaft" (1798), in *Sämtliche Werke* (Munich: Hanser, 1980), 1:352–56.

18. Benjamin Grosser's art project "Facebook Demetricator" does not count either friends or likes: "No longer is the focus on how many friends you have or on how much they like your status, but on who they are and what they said." http://bengrosser.com/projects/facebook-demetricator.

19. For "experience your life," see Gerhard Schulze, *Die Erlebnisgesellschaft. Kultursoziologie der Gegenwart* [The experience society: Cultural sociology of the present] (Frankfurt: Campus, 1992), 58–59. For "narrate yourself," see Dieter Thomä, *Erzähle dich selbst. Lebensgeschichte als philosophisches Problem* [Narrate yourself: Life history as a philosophical problem] (München: C. H. Beck, 1998).

20. Jean M. Twenge, *Generation Me: Why Today's Young Americans Are More Confident, Assertive, Entitled—and More Miserable Than Ever Before* (New York: Free Press, 2006). The relation between depression and social networks is noted in various studies. Examples include Hui-Tzu Chou and Nicholas Edge, "'They Are Happier and Having Better Lives Than I Am': The Impact of Using Facebook on Perceptions of Others' Lives," *Cyberpsychology, Behavior, and Social Networking* 15, no. 2 (2012): 117–21; and Ethan Kross et al., "Facebook Use Predicts Declines in Subjective Well-Being in Young Adults," *PloS ONE 8*, August 14, 2013, http://journals.plos.org/plosone/article?id=10.1371/journal.pone.0069841. On manipulated self-description as mnemotic self-betrayal, see Günter Burkart, "When Privacy Goes Public: New Media and the Transformation of the Culture of Confession," in *Modern Privacy, Shifting Boundaries, New Forms*, ed. Harry Blatterer, Pauline Johnson, and Maria R. Markus, 23–38 (New York: Palgrave McMillan, 2010): "It will become harder for individuals to discern what is their (and their communicating partners') 'real'

self and what is their 'ideal' image they want to present" (35). See also Sarah Knapton, "Lying on Facebook Profiles Can Implant False Memories, Experts Warn," *Telegraph*, December 29, 2014, http:// www.telegraph.co.uk/news/science/science-news/11315319/Lying-on -Facebook-profiles-can-implant-false-memories-experts-warn.html.

21. Dave Eggers's novel *The Circle* (New York: Random House, 2013) gives an exaggerated portrayal of the transparency philosophy of an internet giant called Circle—a mixture of Google and Facebook. The novel (and 2017 movie adaptation) sees itself as the *1984* of the twenty-first century, tearing the mask off the utopia of total transparency and revealing its true face as the dystopia of total surveillance. For Zuckerberg's self-congratulatory praise of the glass architecture of Facebook's headquarters, see www.facebook.com/zuck/videos/10102367711349271. For a discussion of sharing as a "euphemism for selling and commodifying data," see the chapter on Facebook in Christian Fuchs, ed., *Social Media: A Critical Introduction* (London: Sage, 2014), 153–78, esp. 172.

22. Manfred Schneider, *Transparenztraum* [Transparency dream] (Berlin: Matthes und Seitz, 2013). See also Benjamin's praise for "glass-culture," in, among other places, Walter Benjamin, "Experience and Poverty," in *Selected Writings*: vol. 2, part 2: *1931–1934*, ed. Michael W. Jennings, Howard Eiland, and Gary Smith (Cambridge, MA: Harvard University Press, 1999), 734. Among the protagonists of "artistic self-surveillance" is Jennifer Ringley, who from 1996 to 2003, on the website JenniCam, used the camera to make the events in her dorm room and apartment publicly accessible; also Josh Harris, who among other projects published his life with his partner in much the same way. There is a documentary film on Harris by Ondi Timoner titled *We Live in Public* (2009).

23. Peter Singer, "Visible Man: Ethics in a World Without Secrets," *Harper's*, August 2011, http://harpers.org/archive/2011/08/visible-man. See also Christian Heller, *Post-Privacy: Prima leben ohne Privatsphäre* [Post-privacy: Living just fine without a private sphere] (Munich: C. H. Beck, 2011). The argument can be found even before the World Wide Web and September 11, in Joshua Meyrowitz, *No Sense of Place: The Impact of Electronic Media on Social Behavior* (New York: Oxford University Press, 1985), 322.

24. Meyrowitz: *No Sense of Place*, 311. In the final chapter, "Whither '1984,'" Meyrowitz concludes that radio and TV undermine the pyramid model, which has a few people observing and controlling the

broad masses. The reduction of privacy has often been welcomed since then, as a defensive measure against the secrecy of bank accounts and business and income reporting, which in principle protects the rich from social control and public outrage.

25. The new pertinence of old concepts is reflected in the work of Staples, who in 1997 still wrote, "There is no 'Big Brother,' we are him." William G. Staples, *The Culture of Surveillance: Discipline and Social Control in the United States* (New York: St. Martin's, 1997), 129. In a later book, however, he again operates with the concept of Big Brother, in the form of the state, which co-opts diverse private enterprises ("tiny brothers") for the business of surveillance. William G. Staples, *Everyday Surveillance: Vigilance and Visibility in Postmodern Life* (Lanham, MD: Rowman & Littlefield, 2014).

26. Ramón Reichert, "Einführung" [Introduction], in *Big Data. Analysen zum digitalen Wandel von Wissen, Macht und Ökonomie* [Big Data: Analyses of the digital transformation of knowledge, power, and economy], ed. Ramón Reichert (Bielefeld: transcript, 2014), 10. Pertinent to a systematic approach to research from the perspective of critical theory are Fuchs, ed., *Social Media*; and David M. Berry, *Critical Theory and the Digital* (New York: Bloomsbury, 2014). For research on specific topics, see also Martin Kuhn, *Federal Dataveillance: Implications for Constitutional Privacy Protections* (New York: LFB Scholarly Publications, 2007); Christian Fuchs et al., eds., *Internet and Surveillance: The Challenges of Web 2.0 and Social Media* (New York: Routledge, 2012); and Oliver Leistert and Theo Röhle, eds., *Generation Facebook: Über das Leben im Social Net* [Generation Facebook: On life in the social net] (Bielefeld: transcript, 2011). Trebor Scholz, ed., *Digital Labor: The Internet as Playground and Factory* (New York: Routledge, 2013), examines the production of a new class of underpaid day workers in the form of crowdsourcers on the fringes of the political economy of the social media. Jussi Parikka, *A Geology of Media* (Minneapolis: University of Minnesota Press, 2015), sheds light on the materiality of the digital, quite literally in the form of the production of the necessary minerals and their ecological fate after the end of the devices' useful life.

27. On binarizing mathematized communication, see Dieter Mersch, *Ordo ab chao—Order from Noise* (Zurich: Diaphanes, 2013), 47. It should be noted that cybernetic governmentalism is being advanced as a quasi "popular movement" since the controlling knowledge that it

develops is largely provided "from below" in the form of voluntary information about individuals on social networks or as measurement of the self in the context of the Quantified Self movement. It should also be noted that the regime of measurement is, naturally, ideological: The pressure to optimize that is implied and is already betrayed in the name of the app OptimizeMe (meanwhile terminologically optimized to Optimized) aims at the production of "neo-liberal, responsibilized subjectivities" along with the justified fear that the imperative of self-responsibility is ultimately leading to the neoliberal individualization of health services. See Jennifer R. Whitson, "Gaming the Quantified Self," *Surveillance and Society* 11, no. 1/2 (2013): 173; and Melanie Swan, "'Health 2050,' The Realization of Personalized Medicine Through Crowdsourcing, the Quantified Self, and the Participatory Biocitizen," *Journal of Personalized Medicine* 2 (2012): 93–118.

28. For Adorno's verdict on amusement, see the chapter "Culture Industry, Enlightenment as Mass Deception" in Theodor W. Adorno and Max Horkheimer, *Dialectic of Enlightenment*, ed. Gunzelin Schmid Noerr, trans. Edmund Jephcott (Stanford, CA: Stanford University Press, 2002). The comparison with people who are sick is found in Rancière's critique of critical theory, where, in his discussion of the psychology of social critique, he makes the point that "doctors," in order to feel that they themselves are healthy, need "patients" and endlessly reproduce them. Jacques Rancière, "The Misadventures of Critical Thought," in *The Emancipated Spectator*, trans. Gregory Elliott (Brooklyn: Verso, 2009), 47. It is not surprising that the application of critical theory to the digital media repeats the accusation of affirmation and in the process disapproves, for example, of Facebook's "like" button as "[wanting] to spread an affirmative atmosphere." Fuchs, ed., *Social Media*, 160.

29. Eric Schmidt and Jared Cohen, *The New Digital Age: Reshaping the Future of People, Nations, and Business* (New York: Knopf, 2013), 3–4. Mark Zuckerberg declared in October 2010, in an interview with TechCrunch founder Michael Arrington, "People have really gotten comfortable not only sharing more information and different kinds, but more openly and with more people. That social norm is just something that's evolved over time. We view it as our role in the system to constantly be innovating and be updating what our system is to reflect what the current social norms are." Bobbie Johnson, "Privacy No Longer a Social Norm, Says Facebook Founder," *Guardian*, January

11, 2010, http://www.theguardian.com/technology/2010/jan/11/facebook -privacy.

30. "Needed are the formation and composition of future and alternative systems, using civil society movements, public encryption, the democratisation of cryptography, megaleaks and the education of citizens about these systems." Berry, *Critical Theory and the Digital*, 147. On the hope for acts of resistance such as Quit Facebook Day (www.quitfacebookday.com), Web 2.0 Suicide Machine (www .suicidemachine.org), and the nonprofit network diaspora (https:// diasporafoundation.org) as a "socialist Internet project"; see Fuchs, ed., *Social Media*, 173–174. The following quotes on "capitalist company" and the "advertising and economic surveillance machine" are also found there (164, 167).

31. Anders Albrechtslund and Lynsey Dubbeld, "The Plays and Arts of Surveillance: Studying Surveillance as Entertainment," *Surveillance & Society* 3, no. 2/3 (2005): 216–21; Torin Monahan, "Surveillance as Cultural Practice," *Sociological Quarterly* 52 (2011): 495–508.

32. Giorgio Agamben, *Infancy and History: On the Destruction of Experience*, trans. Liz Heron (New York: Verso, 1993), 17. Gumbrecht confirms this observation about flight into the act of photographing in the present ("many tourists today do not really know how to react in the real presence of those monuments that, to see live, they have often invested serious amounts of money") but, like Agamben, confines himself to the simple observation. Hans Ulrich Gumbrecht, *Our Broad Present: Time and Contemporary Culture*, trans. Henry Erik Butler (New York: Columbia University Press, 2014), 17–18. On forms of perception at rock concerts, see http://blog.funkyozzi.com/2011/07 /how-to-choose-smart-phone-at-concert.html.

33. Cited in Roland Barthes, *Camera Lucida: Reflections on Photography*, trans. Richard Howard (New York: Hill and Wang, 1981), 53.

34. The "betrayal" increases with the sheer quantity of images to which we are seduced by digital photography. "The extreme shortening of storage times coupled with simultaneous expansion to near-infinite storage capacity have not led to the past being forgotten; rather, the facility of enjoying the present is its victim. There's no time for that anymore." Siegfried Zielinski, *[. . . After the Media]: News from the Slow-Fading Twentieth Century*, trans. Gloria Custance (Minneapolis, MN: Univocal, 2013), 244. The social network for today's "slide show"

is naturally not only Facebook but also and above all Instagram, which since April 2012 has belonged to Facebook.

35. Pfaller's theory of delegated enjoyment, inspired by Jacques Lacan, would be more plausible if it included other chief witnesses beside the video recorder that "watches" for us the films we never get around to and the copy machine that "reads" for us the academic essay we never study. Robert Pfaller, "The Work of Art That Observes Itself," in *Interpassivity: The Aesthetics of Delegated Enjoyment*, ed. Robert Pfaller (Edinburgh: Edinburgh University Press, 2017), 49–84.

36. For a link between the disappearing capacity for experience among modern humans and their lack of concern for privacy, see Wendy Brown, "'The Subject of Privacy': A Comment on Moira Gatens," in *Privacies: Philosophical Evaluations*, ed. Beate Rössler (Stanford, CA: Stanford University Press, 2004), 140: "If we are subjects increasingly incapable of experience in the Benjaminian and Agambenian sense, might this incapacity be a key to understand our own complicity in an order increasingly indifferent to distinctions between public and private space, and hence private and public experience?" With this brief, unfortunately not further developed comment, the problem of privacy is approached from the perspective of a historical and cultural context rather than as a question of the governmental, normative production of the subject. Agamben, in his essay, refers to Benjamin's "Experience and Poverty."

37. There is no simple way, in English, to mark the difference between German's two words for experience: *Erfahrung* refers to experience gathered over time, with exposure to different events and realities, and implies the acquisition of understanding; *Erlebnis* contains the root of the word "to live" and applies to individual experiences. The editors of Benjamin's *Selected Writings* define the difference as follows: "Benjamin draws . . . on the distinction, developed in the essay 'On Some Motifs in Baudelaire,' between the 'isolated experience' [*Erlebnis*] and the traditional, cohesive, and cumulative experience [*Erfahrung*]." Walter Benjamin, *Selected Writings*, vol. 4: *1938–1940*, ed. Howard Eiland and Michael W. Jennings, trans. Edmund Jephcott and Howard Eiland (Cambridge, MA: Harvard University Press, 2006), 198n68. In the following, *Erfahrung* is generally rendered as "experience" or "long-term experience" and *Erlebnis* as "lived experience." In some instances, the terms are differentiated by the use of additional modifiers.—Trans.

38. Walter Benjamin, "Central Park," in *Selected Writings*, vol. 4: *1938–1940*, ed. Howard Eiland and Michael W. Jennings, trans. Edmund Jephcott and Howard Eiland (Cambridge, MA: Harvard University Press, 2006), 183 (translation modified).

39. Walter Benjamin, "Experience and Poverty," in *Selected Writings*, vol. 2, part 2: *1931–1934*, ed. Michael W. Jennings, Howard Eiland, and Gary Smith (Cambridge, MA: Harvard University Press, 1999), 734.

40. What is described here is the bustling communication on social media, which adheres to a different logic than, say, exclusive sharing (for example, via WhatsApp) with a particular person, who may be expecting the message and who then naturally reacts differently to a photograph (from the faraway city, from a museum) than friends on the network.

41. Douglas Rushkoff, *Present Shock: When Everything Happens Now* (New York: Penguin, 2013), 4. See also Thomas Hylland Eriksen, *Tyranny of the Moment: Fast and Slow Time in the Information Age* (London: Pluto, 2001), 119, who notes, "The moment, or instant, is ephemeral, superficial and intense."

42. This perspective contradicts the positive view of "digital memory items" that José van Dijck, *Mediated Memories in the Digital Age* (Stanford, CA: Stanford University Press, 2007), 48, had expressed, when she called them "networked objects, constructed in the commonality of the World Wide Web in constant interaction with other people." Jacob Silverman offers an account that comes close to my redemptive view of sharing: "Taking photographs gives you something to do; it means that you no longer have to be idle. . . . Living in the moment means trying to capture and possess it." Silverman's presentation of the obsession with photography can be read as a radical delegation of experience: "This kind of cultural practice is no more clearly on display than during a night out with twentysomethings. The evening becomes partitioned into opportunities for photo taking: getting dressed, friends arriving, a taxi ride, arriving at the bar, running into more friends, encountering funny graffiti in the bathroom, drunk street food, the stranger vomiting on the street, the taxi home, maybe a shot of the clock before bed. A story is told here, sure, but more precisely, life is documented, its reality confirmed by being spliced into shareable data. Now everyone knows how much fun you had and offers their approval, and you can return to it to see what you forgot in that boozy haze." Jacob Silverman, "'Pics or It Didn't

Happen'—The Mantra of the Instagram Era: How Sharing Our Every Moment on Social Media Became the New Living," *Guardian*, February 26, 2015, http://www.theguardian.com/news/2015/feb/26/pics-or-it-didnt-happen-mantra-instagram-era-facebook-twitter. It does not require a lot of insight to look beyond the argument of narcissistic self-representation and see, in the obsession with capturing experiences photographically that is described here, a profound melancholy vis-à-vis the impending pastness of the present—a prospective nostalgia that, instead of an individual really living "in the moment" and letting it pass by lightheartedly, actually makes that individual the victim of the ineluctable transience of life—a *memento mori* that exceeds the nostalgia Sontag ascribes to photography, since it does not refer to "another person's (or thing's) mortality" but to that of the photographer herself. Susan Sontag, *On Photography* (New York: Picador, 1973), 11.

43. That Facebook actualizes and radicalizes this aspect is at the heart of the present argument, which is indirectly supported by the mashup in a scene from the TV series *Mad Men*, in which Don Draper's proposal for the Kodak carousel is transformed into a Facebook timeline (http://www.youtube.com/watch?v=r6Th2omR8UI). Facebook's photo app Moments, which is advertised with the slogan "Get the photos you didn't take," is a kind of retrieval of moments in advance of the camera, at least of one's own camera, because it uses facial recognition and automatic assignment of names to make it possible to collect the photos others have taken of you. What from the perspective described here could be understood as a recovery of the experienced moment is, in the eyes of the data protectors (who protested against this app in the summer of 2015 in Europe), naturally only one more step in the all-encompassing loss of privacy.

44. To an extent, the perspective presented here refers back to Christopher Lasch's view, according to which the culture of narcissism of the 1970s and 1980s is not an expression of egoism and selfishness but "a culture of survivalism." "Narcissism signifies a loss of selfhood, not self-assertion. It refers to a self threatened with disintegration and by a sense of inner emptiness. Christopher Lasch, *The Minimal Self: Psychic Survival in Troubled Times* (New York: Norton, 1984), 57. However, the justification for inner emptiness is now found not in the "doomsday mentality" (93) of the Cold War period and ecological apocalypse but in the lack of prospects and meaning accorded to the

individual's own existence. What must be survived—or, as Lasch says, "coped with" —is less the threat to life (which, in actuality, has not been ecologically or politically diminished) than its banality. On the betrayal of the present in the interest of the future, compare Zielinski: "Yet to already be the subject of a past event in the instant that something happens is tantamount to abolishing the present. The present becomes merely an extremely short effect for the future." Zielinski, [. . . After the Media], 244.

45. On the evidence of a lack of need for solitude, see William Deresiewicz, "The End of Solitude," Chronicle of Higher Education, January 30, 2009, http://chronicle.com/article/The-End-of-Solitude/3708. On the joys of solitude, compare Turkle, Alone Together, 27. On social networks as a "distraction from the torture of now-time," see Geert Lovink, Ippolita, and Ned Rossiter, "The Digital Given: 10 Web 2.0 Theses," Fibreculture Journal 14 (2009), http://fourteen.fibreculturejournal.org /fcj-096-the-digital-given-10-web-2-0-theses.

46. Max Picard, The World of Silence, trans. Stanley Godman (Chicago: Henry Regnery, 1952). For the warning about "Radioitis," see Friedrich Pütz, "Die richtige Diät des Hörers" [The proper diet for the listener] (1927), in Medientheorie. 1888–1933. Texte und Kommentare [Media theory, 1888–1933: Texts and commentaries], ed. Albert Kümmel and Petra Löffler (Frankfurt: Suhrkamp, 2002), 275. On September 27, 2015, in a speech to bishops in Pennsylvania, Pope Francis complained that social media inhibit the creation of real relationships and make people lonely. "Social bonds are a mere means for satisfaction of my needs. . . . I would dare say that at the root of so many contemporary situations is a kind of radical loneliness that so many people live in today. Running after the latest fad, a like, accumulating followers on any of the social networks. And we human beings get caught up in what contemporary society has to offer: loneliness with fear of commitment." http://www.wired.co.uk/news/archive/2015-09 /28/pope-francis-social-media-causes-loneliness. The Wired release also contains the video clip with the English translation that is given here. Francis's predecessor, Pope Benedict, in his message to the Forty-Sixth World Communication Day, May 20, 2012, "Silence and Word: Path of Evangelization," had already problematized the internet as a danger for "that silence which becomes contemplation, which introduces us into God's silence." http://w2.vatican.va/content/benedict

-xvi/en/messages/communications/documents/hf_ben-xvi_
mes_20120124_46th-world-communications-day.html.

47. Blaise Pascal, Pensées no. 139, in *Opuscules et pensées* (Paris: Hachette, 1897).

48. Gilles Deleuze, "Mediators," in *Negotiations 1972–1990*, trans. Martin Joughin (New York: Columbia University Press, 1997), 129.

49. Picard, *The World of Silence*, 221. Long before the neologism "produser" (from producer and user), Michael Joyce came up with the concept of the "wreader," from reader and writer, which was symptomatic of the then popular celebration of the liberation of the reader from domination by the author, which was criticized at the time but has meanwhile turned out to be the incapacity to engage in concentrated reading or listening. Michael Joyce, "Nonce Upon Some Times: Rereading Hypertext Fiction," *Modern Fiction Studies* 43, no. 3 (1997): 579–97.

50. Antoine de Saint-Exupéry, "Flight to Arras," in *Airman's Odyssey*, trans. Lewis Galantière (New York: Reynal & Hitchcock, 1942), 346–47.

51. Friedrich Daniel Ernst Schleiermacher, "Versuch einer Theorie des geselligen Betragens" [Attempt at a theory of social behavior], in *Monologen: Eine Neujahrsgabe* [Monologues: A New Year's gift] (Berlin: Holzinger, 2016).

52. Johann Wolfgang von Goethe, *Faust*, trans. Martin Greenberg (New Haven, CT: Yale University Press, 1992, 1998), part 1, verses 1718–20.

53. Goethe, *Faust*, part 2, verse 4936.

54. Gotthold Ephraim Lessing, *The Education of the Human Race* (London: Smith, Elder & Co., 1858). Immanuel Kant conceived the Enlightenment project in his 1784 essay "Answer to the Question: What is Enlightenment" as "man's emergence from his self-imposed nonage." In this connection, he speaks in the *Groundwork for the Metaphysics of Morals* (1797) of a "duty of man toward himself." Immanuel Kant, "An Answer to the Question: 'What Is Enlightenment?'" trans. H. B. Nisbet (London: Penguin, 2013); Immanuel Kant, *Groundwork of the Metaphysics of Morals*, trans. Mary Gregor and Jens Timmermann (Cambridge: Cambridge University Press, 2012).

55. Ludwig Tieck, *William Lovell*, trans. Douglas Robertson (2009), part 4, book 6, no. 11, William to Rosa (29).

56. Bloch, in this sense, sees in Goethe's Faust "the highest example of utopian man." Ernest Bloch, *The Principle of Hope*, trans. Neville Plaice, Stephen Plaice, and Paul Knight (Cambridge, MA: MIT Press,

1986), 1012. The reference is thanks to Bruno Hillebrand's study *Ästhetik des Augenblicks. Der Dichter als Überwinder der Zeit—von Goethe bis heute* [Aesthetic of the moment: The poet as victor over time—from Goethe to the present] (Göttingen: Vandenhoeck & Ruprecht, 1999).

57. On the fateful triumph of *homo faber*, who transforms mankind into "the compulsive executor of his capacity," see Hans Jonas, *The Imperative of Responsibility: In Search of an Ethics for the Technological Age*, trans. Hans Jonas and David Herr (Chicago: University of Chicago Press, 1984), 142. Similarly, Nancy refers to the changed nature of man's desire for conquest as "no longer the domination by the 'bourgeois' but by the machine they had served" and calls for exiting from a teleology that has "its own ends, indifferent to the existence of the world and of all its beings." Jean-Luc Nancy, *After Fukushima: The Equivalence of Catastrophes*, trans. Charlotte Mandell (New York: Fordham University Press, 2014), 7, 12.

58. On the end of grand narratives (*grands récits*), see Jean-François Lyotard, *The Postmodern Condition: A Report on Knowledge*, trans. Geoff Bennington and Brian Massumi (Minnesota: University of Minnesota Press, 1984). On the end of history (which by no means signifies the end of historical events or social problems) in the mode of liberal democracy, which reveals and surmounts all cultural differences as phenomena derived from different phases of historical development, see Francis Fukuyama: *The End of History and the Last Man* (New York: Avon, 1993).

59. Zygmunt Bauman, "From Pilgrim to Tourist—or a Short History of Identity," in *Questions of Cultural Identity*, ed. Stuart Hall and Paul du Gay (London: Sage, 1996), 24. On the analogy of the raftsmen and the sailors, see Bauman, *Liquid Life* (Cambridge, MA: Polity, 2005), 20.

60. Zygmunt Bauman, "Privacy, Secrecy, Intimacy, Human Bonds, Utopia—and Other Collateral Casualties of Liquid Modernity," in *Modern Privacy: Shifting Boundaries, New Forms*, ed. Harry Blatterer, Pauline Johnson, and Maria R. Markus (New York: Palgrave Macmillan, 2010), 19–20.

61. Nathan Jurgenson, "The Facebook Eye," *Atlantic*, January 12, 2012, http://www.theatlantic.com/technology/archive/2012/01/the-facebook-eye/251377.

62. Bauman comments: "Unlike the utopias of yore, the hunters' utopia does not offer a meaning to life—whether genuine or fraudulent. It

only helps to chase the question of life's meaning away from the mind of living. Having reshaped the course of life into an unending series of self-focused pursuits and with each episode lived through as an overture to the next, it offers no occasion for reflection about the direction and the sense of it at all." Also of note is Bauman's comment on "the end of time as *history*" and the ironically presumed utopian status of the hunter society: "Strange, unorthodox utopia it is—but utopia all the same, as it promises the same unattainable prize all utopias brandished, namely the ultimate and radical solution to human problems past, present and future, and the ultimate and radical cure for the sorrows and pains of the human condition." Baumann, "Privacy," 22, 21.

63. Tiqqun, *Theory of Bloom*, trans. Robert Hurley (London: LBC, 2012), 44, 63.

64. Tiqqun, *Theory of Bloom*, 16, 19, 59, 21, 79, 65.

65. Jacques Rancière, *Aesthetics and Its Discontents*, trans. Steven Corcoran (New York: Polity, 2009), 104.

66. Tiqqun, *Theory of Bloom*, 105.

67. Jean-Luc Nancy, *Die herausgeforderte Gemeinschaft* [The challenged community], trans. (German) Esther von der Osten (Berlin: Diaphanes, 2007), 28. The essay appeared as the foreword to the Italian edition of Maurice Blanchot's *The Unavowable Community.*—Trans.

68. Zielinski, [. . . *After the Media*], 249.

69. Alexander Pschera, *800 Millionen. Eine Apologie der sozialen Medien* [800 million: An apologia for social media] (Berlin: Matthes & Seitz, 2011), 43, 62, 45. "The network, as 'social,' provides a speakable universal language and offers the human community new possibilities for understanding and rapprochement regarding the utopian project toward whose realization we, as social beings, are continuously striving" (24). Lovink, taking the opposing view, speaks of a "culture of 'detached engagement'" that lacks socially and politically relevant goals and concepts. Geert Lovink, *Networks Without a Cause: A Critique of Social Media* (Malden, MA: Polity, 2012), 2. His accusation of time wasting—"the networks without cause are time eaters, and we're only being sucked deeper into the social cave without knowing what to look for"—elicits a cryptic but definitive response from Pschera, who writes (without mentioning Lovink), "those who accuse the social media of ornamentalizing are only making a taboo of their potential to tunnel under things" (61). The following quotations are from pages 65 and 66.

70. On phatic communication on social networks, see Miller: "Communication has been subordinated to the role of the simple maintenance of ever expanding networks and the notion of a connected presence." Vincent Miller, "New Media, Networking, and Phatic Culture," *Convergence 14* (2008): 398. The notion of a communication utopia without actual things that are communicated also links to Agamben, who characterizes children's experience of linguistic capacity as "experience of language as such, in its pure auto-reference," as experience "of the pure fact that one speaks, that language exists." Giorgio Agamben, *Infancy and History: On the Destruction of Experience*, trans. Liz Heron (New York: Verso, 1993), 6.

71. As an example, see part V, "There is much metaphysics in thinking of nothing," in Caeiro's poetry collection *O Guardador de Rebanhos*. Eduardo Caeiro and Fernando Pessoa, *The Keeper of Sheep*, trans. Edwin Honig and Susan M. Brown (Bronx, NY: Sheep Meadow, 1997).

72. Hans Ulrich Gumbrecht, *Production of Presence: What Meaning Cannot Convey* (Stanford, CA: Stanford University Press, 2004), 79, 117. The critique of meaning culture occurs both previous and parallel to Gumbrecht's intervention on various other discursive fronts. In this connection, the following texts are also of interest: Susan Sontag, *Against Interpretation* (New York: Farrar, Straus and Giroux, 1966); Jochen Hörisch, *Wut des Verstehens: Zur Kritik der Hermeneutik* [The rage of understanding: Toward a critique of hermeneutics] (Frankfurt: Suhrkamp, 1988); Jean-Luc Nancy, *The Birth to Presence*, trans. Brian Holmes et al. (Stanford, CA: Stanford University Press, 1993); Erika Fischer-Lichte, *The Transformative Power of Performance: A New Aesthetic*, trans. Saskya Iris Jain (New York: Routledge, 2004). I discuss the connections and difference of these positions in my study *Digital Art and Meaning: Reading Kinetic Poetry, Text Machines, Mapping Art, and Interactive Installations* (Minneapolis: University of Minnesota Press, 2011), from which some of the ideas presented here are derived.

73. Gumbrecht, *Production of Presence*, 145, 138. For a further illustration of Gumbrecht's fatalistic view of history, which dismisses "historical thinking" and presents time as an "agent of change" and the future as an "open horizon of possibilities," see Gumbrecht, *Our Broad Present*, 14.

74. Theodor W. Adorno, *Minima Moralia: Reflections from Damaged Life*, trans. Edmund Jephcott (New York: Verso, 2006), 25.

75. Hans Ulrich Gumbrecht, *In 1926: Living at the Edge of Time* (Cambridge, MA: Harvard University Press, 1997), 186. What is at stake here is not experience as an epistemological problem in the sense of Immanuel Kant, or the further transformation of his categories of cognition into factors based on social schemata, but differing levels of intensity in perception of the world. For the present discussion of experience, this perspective is determinative, with a central role assigned to media-specific dispositions (the significance of the newspaper as addressed by Gumbrecht, the function of photography as addressed by Agamben, and various discussions of hyper-reading in the internet).

76. Gumbrecht, *In 1926*, 187.

77. Gumbrecht, *In 1926*, 188. The Egon Erwin Kisch quotation is taken from his *Hetzjagd durch die Zeit* [Feverish hunt through time] (Berlin: Universum Bücherei, 1926). Kisch's collection of reportage *Der rasende Reporter* [The raging reporter] appeared in 1925.

78. See Helmut Lethen, *Cool Conduct: The Culture of Distance in Weimar Germany*, trans. Don Reneau (Berkeley: University of California Press, 2002). The term "cool," here, is used in a way that implies coldness and not mere stylistic sophistication.—Trans.

79. Kurt Pinthus, "Masculine Literature" (1929), in *The Weimar Republic Sourcebook*, ed. Anton Kaes, Martin Jay, and Edward Dimendberg (Berkeley: University of California Press, 1994), 519.

80. Ernst Jünger on photography, cited by Lethen, *Cool Conduct*, 148.

81. Bertolt Brecht, *Bertolt Brecht on Film and Radio*, ed. and trans. Marc Silberman (London: Bloomsbury, 2000), 144; Jean Baudrillard, *Photographies 1985–1998* [exhibition catalog], ed. Peter Weibel, trans. Susanne Baumann et al. (Ostfildern-Ruit: Hatje Cantz, 1999), 24–25, 28. The understanding of photography as a cold medium, as presented here, differs from Marshall McLuhan's distinction between cold and hot media based on their sensual quality and wealth of detail, according to which photography is a hot medium because (unlike "cold" caricature) it is rich in optical detail. Marshall McLuhan, *Understanding Media: The Extensions of Man* (Cambridge, MA: MIT Press, 1994), 22–23.

82. Baudrillard, *Photographies 1985–1998*, 24, 22.

83. Béla Balász, *Early Film Theory: Visible Man and the Spirit of Film*, trans. Rodney Livingstone (New York: Berghahn, 2011), 157.

84. Walter Benjamin, "On Some Motifs in Baudelaire," in *Selected Writings*, vol. 4: *1938–1940*, ed. Howard Eiland and Michael W. Jennings,

trans. Harry Zohn (Cambridge, MA: Harvard University Press, 2006), 315–16.

85. Walter Benjamin, *The Arcades Project*, trans. Howard Eiland and Kevin McLaughlin (Cambridge, MA: Harvard University Press, 2002), 801. Shortly after this remark, Benjamin defines leisure as "an early form of distraction or amusement" (804, m4, 1).

86. Benjamin, "Experience and Poverty," 734.

87. Benjamin, "Experience," 3.

88. Benjamin, *Arcades Project*, 473 (N9a, 1) (translation modified).

89. Walter Benjamin, "On the Concept of History," in *Selected Writings*, vol. 4: *1938–1940*, ed. Howard Eiland and Michael W. Jennings, trans. Harry Zohn (Cambridge, MA: Harvard University Press, 2006), 390 (translation modified). The dialectical image is "an image that emerges suddenly, in a flash," and shows the past not the way it supposedly "actually was" but in its hidden constellation and relation to the present. Benjamin, *Arcades Project*, 473 (N9, 7). On Benjamin's critique of historicism's purportedly "objective" history writing, which in Benjamin's eyes basically conformed to the perspective of the ruling class as "the strongest narcotic of the century" (463; N3, 4), see his "Fragmente zur Sprachphilosophie und Erkenntniskritik" [Fragments on language philosophy and epistemology], in *Kairos. Schriften zur Philosophie*, ed. Ralf Konersmann (Frankfurt: Suhrkamp, 2007), 75.

90. Ralf Konersmann, "Nachwort. Walter Benjamins philosophische Kairologie" [Afterword: Walter Benjamin's philosophical kairology], in *Kairos. Schriften zur Philosophie*, 334. The trick in Benjamin's argument is the implication of a kind of "depth" photography that puts the optical unconscious to historico-philosophical use, as details that were drowned in the flow of events, but that the photograph, by bringing time to a halt and magnifying the event, makes conscious again. In this sense, Benjamin can also speak of an "intentionless truth" (of the object itself). This fundamentally metaphysical impulse makes it possible to escape "spiritless" life by means of a new "grand" plan. It simultaneously runs the risk of replacing the old certainties by new, politically justified ones. On the critique of the metaphysical presuppositions of thought in Benjamin, see Karl Heinz Bohrer, *Der Abschied: Theorie der Trauer: Baudelaire, Goethe, Nietzsche, Benjamin* [The parting: Theory of mourning: Baudelaire, Goethe, Nietzsche, Benjamin] (Frankfurt: Suhrkamp, 1996).

91. Tiqqun, *Theory of Bloom*, 37, 38.
92. Siegfried Kracauer, *The Salaried Masses: Duty and Distraction in Weimar Germany*, trans. Quintin Hoar (London: Verso, 1998), 88.
93. Benjamin, "Central Park," 183.
94. Benjamin, "Experience and Poverty," 734.
95. There have been repeated calls for deceleration from academics. See, for example, Thomas Hylland Eriksen, *Tyranny of the Moment: Fast and Slow Time in the Information Age* (London: Pluto, 2001), which ends with a chapter on "The Pleasures of Slow Time"; Byung-Chul Han, *The Scent of Time: A Philosophical Essay on the Art of Lingering*, trans. Daniel Steuer (New York: Polity, 2017); Hartmut Rosa, *Acceleration and Alienation: Toward a Critical Theory of Late Modern Temporality* (Aarhus: Aarhus University Press, 2010); and Lutz Koepnick, *On Slowness: Toward an Aesthetic of the Contemporary* (New York: Columbia University Press, 2014). Popular science books on the subject include Karlheinz Geissler's *Lob der Pause. Von der Vielfalt der Zeiten und der Poesie des Augenblicks* [Praise of pauses. On the diversity of times and the poetry of the moment] (Munich: Oekom, 2012); Eduard Kaiser, *Trost der Langeweile: Die Entdeckung menschlicher Lebensformen in digitalen Welten* [The consolation of boredom: The discovery of human life forms in digital worlds] (Rüegger, 2014); and Pico Iyer, *The Art of Stillness—Adventures in Going Nowhere* (New York: Simon & Schuster, 2014). There are films, for example, Florian Opitz's *Speed—Auf der Suche nach der verlorenen Zeit* [Speed—In search of lost time] (2012), along with various other variations on the concept of slowness: slow TV, which, in the tradition of Andy Warhol's *Sleep*, presents banal everyday events in unabbreviated form (the 134-hour journey of a ship from Bergen, Norway, to Kirkenes, in June 2011, which was shown in its entirety on Norwegian television); slow publishing (the British magazine *Delayed Gratification*, which reports on events that occurred at least three months earlier); or slow food. An example of a voluntary attempt to resist the lure of social media was the only partly successful attempt of a group of fifteen-year-olds in London, early in 2015, to stay off of social media for a week. As the group's report details, some of them eventually got around to reading books, while others didn't know what to do with the time that had been freed up: http://www.bbc.co.uk/schoolreport/31942696.

2. AUTOMATIC AUTOBIOGRAPHY

1. Arguments like these actually get published. Kucklick notes that in the United States every day more than 16 billion words are published on Facebook alone, something completely unprecedented in history, since until the end of the twentieth century few people wrote for private reasons. Christoph Kucklick, *Die granulare Gesellschaft: Wie das Digitale unsere Wirklichkeit auflöst* [Granular Society: How the digital dissolves our reality] (Berlin: Ullstein, 2014), 229–30. His overhasty conclusion: "If it is correct that writing serves self-knowledge, then we are experiencing an intensification of sensibilities in regard to ourselves" (231).

2. Nancy K. Miller, "The Entangled Self: Genre Bondage in the Age of the Memoir," *PMLA* 122, no. 2 (2007): 545.

3. Shanyang Zhao, Sherry Grassmuck, and Jason Martin, "Identity Construction on Facebook: Digital Empowerment in Anchored Relationships," *Computers in Human Behavior* 24, no. 5 (2008): 1816–36. As van Dijck remarks, "'liking' has turned into a provoked automated gesture that yields precious information about people's desires and predilections." José van Dijck, "'You Have One Identity': Performing the Self on Facebook and LinkedIn," *Media, Culture & Society* 35, no. 2 (2013): 202.

4. In Germany many sites restrict the automatism of so-called social plug-ins via the double-click principle, which requires a first click to activate the link to Facebook, by which users consciously confirm the data transfer. However, even the double-click solution often allows the link to be permanently activated, whereas on non-German or non-European websites the link to Facebook is usually established without users' knowledge.

5. Samuel Gibbs, "Facebook Tracks All Visitors, Breaching EU Law," *Guardian*, March 31, 2015, http://www.theguardian.com/technology /2015/mar/31/facebook-tracks-all-visitors-breaching-eu-law-report.

6. Hayden White, *The Content of Form: Narrative Discourse and Historical Representation* (Baltimore, MD: Johns Hopkins University Press, 1987), 6, 11, 8.

7. Adam Weishaupt presented his *Geschichte der Vervollkommnung des menschlichen Geschlechts* [History of the perfection of the human race], commencing in 1788, as a "history without years and names." Cited by Reinhart Koselleck and Horst Günther, "Geschichte" [History], in

Geschichtliche Grundbegriffe. Historisches Lexikon zur politisch-sozialen Sprache in Deutschland [Basic historical concepts: Historical lexicon of political and social language in Germany], ed. Otto Brunner, Werner Conze, and Reinhart Koselleck (Stuttgart: Klett-Cotta, 1975), 2:651. The "law of historical conservation of energy" was formulated by Johann Gottfried Herder in 1774: "As, now, since the creation of our earth no ray of sunlight has been lost on her, so also no fallen leaf of a tree, no wind-blown seed of a plant, no corpse of a rotting animal, much less an action of a living being, has remained without effect." *Herders Sämtliche Werke*, ed. Bernhard Suphan (Berlin: Weidmannsche Buchhandlung, 1877–1913), 14:236. Kant prominently issued the call for a historian who would discover a deeper "natural purpose in this idiotic course of things human" as the a priori of a philosophical chiliasmus. Immanuel Kant, "Idea for a Universal History with a Cosmopolitan Purpose," in *On History*, trans. Lewis White Beck (Indianapolis, IN: Bobbs-Merrill, 1963), 12.

8. Wilhelm von Humboldt, "On the Historian's Task [1821]," *History and Theory* 6, no. 1 (1967): 58.

9. Pierre Nora, "Between Memory and History: Les Lieux de Mémoire," *Representations* 26 (Spring 1989): 18.

10. Cited according to Koselleck and Günther, "Geschichte," 663.

11. Charles S. Peirce, *Phänomen und Logik der Zeichen* (Frankfurt: Suhrkamp, 1983), 65. The indexicality of photographs already does not apply to early art photography or to the politically or aesthetically motivated retouching of analog images. It has become thoroughly dubious given the possibilities for manipulation offered by digital photography. However, the indexical paradigm is quite suitable as a descriptive category for the normal case of private photos and as a metaphor for the discussion of "photographic writing" that we intend here. The critique of "daguerrotypical" realism as idolatry derives from the literary scholar Robert E. Prutz, 1856, in *Deutsches Museum. Zeitschrift für Literatur, Kunst und öffentliches Leben* [German museum: Journal for literature, art, and public life], cited by Ulf Eisele, "Empiristischer Realismus. Die epistemologische Problematik einer literarischen Konzeption" [Empiristical realism: The epistemological problematic of a literary conception], in *Naturalismus, Fin de siècle, Expressionismus. 1890–1918* [Naturalism, fin de siècle, Expressionism: 1890–1918], ed. York-Gothart Mix (Munich: Hanser, 1996), 78, 76. This critique would be more applicable to naturalism (and even then

only with regard to theoretical propositions) if the latter, under the impression of the then dominant theory of social determinism, were to replace the productive imagination of an author with an objective description (i.e., a kind of "nonmechanical" mechanical reproduction). See Wilhelm Bölsche, *Die naturwissenschaftlichen Grundlagen der Poesie. Prolegomena einer realistischen Ästhetik* [The natural-scientific foundations of poetry: Prolegomena to a realistic aesthetics] (Leipzig: Karl Reissner, 1887).

12. Siegfried Kracauer, "Photography," *Critical Inquiry* 19, no. 3 (Spring 1993): 425. The following sentence reads: "Since what is significant is not reducible to either merely spatial or merely temporal terms, memory-images are at odds with photographic representation. From the latter's perspective, memory-images appear to be fragments but only because photography does not encompass the meaning to which they refer and in relation to which they cease to be fragments. Similarly, from the perspective of memory, photography appears as a jumble that consists partly of garbage."

13. Baudrillard, *Photographies 1985–1998* [exhibition catalog], ed. Peter Weibel, trans. Susanne Baumann et al. (Ostfildern-Ruit: Hatje Cantz, 1999), 24–25.

14. "The last image of a person is that person's actual *history*." Kracauer, "Photography," 426. Paul Ricœur, *Time and Narrative*, trans. Kathleen McLaughlin and David Pellauer (Chicago: University of Chicago Press, 1984), 3. Narration is understood here, in general terms, as the difference between an initial and an end state, with an at least chronological, if not causal order of events between the two poles. The description usually starts from the endpoint of the narrative (closed narration), unless the temporal position of the narrative is itself the endpoint (open narrative). As argued in this chapter, the narrower, *verbal* understanding of narrative dominates, in order to emphasize the contrast with a visual (and in principle less conscious) representation of reality.

15. Jerome S. Bruner, "Past and Present as Narrative Constructions," in *Narration, Identity, and Historical Consciousness*, ed. Jürgen Straub (New York: Berghahn, 2005), 26. Polkinghorne defines the function of narrative as follows: "Narrative is the cognitive process that gives meaning to temporal events by identifying them as parts of a plot." Donald E. Polkinghorne, "Narrative and Self-Concept," *Journal of Narrative and Life History* 1, no. 2/3 (1991): 136. See also Eakin: "When

it comes to our identities, narrative is not merely about self, but is rather in some profound way a constituent part of self." Paul John Eakin, *Living Autobiographically: How We Create Identity in Narrative* (Ithaca, NY: Cornell University Press, 2008), 2. On autobiography not as reconstruction, but as construction of the self, see also Paul de Man, "Autobiography as De-Facement," *MLN* 94, no. 5 (December 1979): 919–30.

16. Hartmut Rosa, *Weltbeziehungen im Zeitalter der Beschleunigung: Umrisse einer neuen Gesellschaftskritik* [Relations to the world in the era of acceleration: Outlines of a new social critique] (Berlin: Suhrkamp, 2012), 225.

17. Janis Forman, *Storytelling in Business: The Authentic and Fluent Organization* (Stanford, CA: Stanford University, 2013); Frederick W. Mayer, *Narrative Politics: Stories and Collective Action* (Oxford: Oxford University Press, 2014); Rita Charon, *Narrative Medicine: Honoring the Stories of Illness* (Oxford: Oxford University Press, 2006); Barbara Czarniawska, *Narratives in Social Science Research* (London: Sage, 2004); Bernd Kracke and Marc Ries, eds., *Expanded Narration. Das Neue Erzählen* [Expanded narration: The new story-telling] (Bielefeld: transcript, 2013); Hanna Meretoja, *The Narrative Turn in Fiction and Theory: The Crisis and Return of Storytelling from Robbe-Grillet to Tournier* (New York: Palgrave Macmillan, 2014).

18. Lilie Chouliaraki, *The Ironic Spectator: Solidarity in the Age of Posthumanitarianism* (Malden, MA: Polity, 2013), 164. See 167 on the "increasing dispersion of the narrative structure of the news towards testimonial and participatory performances of witnessing" and the "logic of the news as a 'database.'"

19. In this context we should mention StoryCorps, a project founded in 2003 to encourage storytelling. In the tradition of oral history, it makes it possible for all kinds of people, one on one, to have a forty-minute-long conversation in a specially created StoryBooth installed in a public place. The conversations are recorded, given to the participants in the form of a CD, and (with their agreement) archived in the American Folklife Center of the Library of Congress, where they are open to the public. Dave Isay, its founder, in a March 2015 TED talk entitled "Everyone Around You Has a Story the World Needs to Hear" (http://www.ted.com/talks/dave_isay_everyone_around_you _has_a_story_the_world_needs_to_hear), gave an impressive report on the meaning these conversations have for the self-understanding

of the speakers. In his TED talk, Isay also promoted the StoryCorps app, which allows people to record conversations independently of the StoryBooths and StoryCorps staff; he hopes it will lead to 100,000 or more conversations per year. The result can be found on the Story-Corps website, which in May 2015 included approximately 57,000 interviews since 2003. In July 2017, "more than 65,000 interviews" were reported, which suggests an annual average of fewer than 5,000 interviews. The proposal to use a "national homework assignment" to encourage high-school students to engage in conversations with their grandparents or other important people in their lives could raise the number significantly. The question, admittedly, is whether Story-Corps is able to prompt such homework assignments and is able to succeed without them.

20. Reichert notes: "To be entered into the format of the e-questionnaire, linear and narrative knowledge must be broken down into blocks of information. These rules, which are inherent in the form, establish the authority of the e-questionnaire." The "authority" of forms as "hierarchical frames for relationality" lies between the questionnaire and the user as source of information. Ramón Reichert, *Die Macht der Vielen. Über den neuen Kult der digitalen Vernetzung* [The power of the many: On the new cult of digital networking] (Bielefeld: transcript, 2013), 60, 61. That the framework provided must be more politically correct than individual statements is shown by the #FatIsNotaFeeling petition that was launched in early 2015 against the emoticon "feeling fat," which Facebook had offered as a status update. Facebook design-ers are not allowed to make something available just because Facebook users might want to write it. Since Facebook wishes to maintain the listing, which harmonizes nicely with the database, and since formu-lating an expression of emotion was not itself the bone of contention, "feeling 'fat'" was replaced in March 2015, in response to the initiative of the Endangered Bodies Organization, by "feeling stuffed."

21. Page exemplifies the narrative sequences of "small stories" with the help of three status updates by a Facebook user named Cheryl: "Cheryl is a cake lover!" (May 20, 7:29 p.m.), "Cheryl is giving up the cake . . . as of tomorrow!!" (May 21, 9:21 p.m.), "Cheryl did not eat any cake today . . . result!!" (May 22, 6:10 p.m.). Ruth Page, "Reexamin-ing Narrativity: Small Stories in Status Updates," *Text and Talk* 30, no. 4 (2010): 433; 437 on the completion of plot lines by "friends. See also Reichert, *Die Macht der Vielen*: "The personal information, status

updates and comments are . . . largely enumerative. They cite data, accumulate found information and only seldom offer coherent narratives" (61).

22. Status updates like "Francis is in Starbucks" or "Joanna is at home" (Page, "Reexamining Narrativity," 432–33) recall the uncommented listing of events that we saw in the *Annals*—"listings that are not themselves the event but only remind us of the latter: there, the passing of time; here, the existence of the reporter." Page's own findings that "recency is prized over retrospection" and that "these sequences are 'mere successions of doings,' rather than exhibiting the tightly knit, interdependent connections required of narrative sequences in its strictest sense" (440, 439) undermine her proposal to read the individual status updates, in light of Ricœur, as an "attempt to 'make time human' by selecting particular events as worthy of narration" (428). On "pointillist technique" instead of "linear connections between individual entries," see 440; on "pointillist time," see Zygmunt Bauman, "Privacy, Secrecy, Intimacy, Human Bonds, Utopia—and Other Collateral Casualties of Liquid Modernity," in *Modern Privacy: Shifting Boundaries, New Forms*, ed. Harry Blatterer, Pauline Johnson, and Maria R. Markus (New York: Palgrave Macmillan, 2010), 21. Facebook users, in Ricœur's terms, do not approach the second level of mimesis—narrative configuration under the sign of "emplotment"—but remain, if at all, at the level of prefiguration, as a reporting and value-ascribing preconception of what may be taking place.

23. These categories include, among others, "Connections" (links to pages that the person "liked"), "Events" (activities to which she or he was invited), "Wallposts" (postings by other people on the person's bulletin board), "Shares" (all links posted on the bulletin board), "Pokes" (all nudges that have been sent or received), and "Friend Request" (invitations to "befriend" a person, with date and a note if it was refused). For details, see the website *Europe Versus Facebook*: http://europe-v -facebook.org/DE/Datenbestand/datenbestand.html.

24. The welcome side effect, if indeed it is not the actual purpose, of algorithmic storytelling aids is disciplining Facebook's users to be precise in describing their images, since only images with "proper" tags ("Eiffel Tower," not "Iron Phallus"; "Hans Martin" rather than "a good friend") can be accounted for by the service. The storyteller is the "filler" at the front end of the interface, who is present in order to provide more reliable data to the back end.

25. William Davis, "Mark Zuckerberg and the End of Language," *Atlantic*, September 11, 2015, http://www.theatlantic.com/technology/archive/2015/09/silicon-valley-telepathy-wearables/404641. Compare the report by Stuart Dredge, "Facebook Boss Mark Zuckerberg Thinks Telepathy Tech Is on Its Way," *Guardian*, July 1, 2015, http://www.theguardian.com/technology/2015/jul/01/facebook-mark-zuckerberg-telepathy-tech. On the future of Oculus Rift, see Zuckerberg's Facebook page of March 25, 2014: http://www.facebook.com/zuck/posts/10101319050523971.

26. The central motif behind the concept of unconscious sharing is without doubt the notion that it simultaneously produces important information for personalized marketing campaigns and "predictive shopping." It is also true that Zuckerberg is also thinking of other domains when, in response to a question about Facebook's future in journalism, he hopes to see "more immersive content like VR" on Facebook and compares this "rich content" favorably to "just text and photos." The BBC's slogan "We don't just report a story, we live it" could thus turn out to be true in an unexpected and unwanted way, if eyewitnesses with immersive material about "developing stories" publicly challenge the laborious research of reporters. "Townhall Q&A," July 1, 2015, on Zuckerberg's Facebook page: http://www.facebook.com/zuck/posts/10102213601037571.

27. Diverse screenshot apps and Snapchat Inc.'s possible access to the deleted photos quickly led to accusations of false advertising by the Federal Trade Commission, as shown by their press release of May 8, 2014: http://www.ftc.gov/news-events/press-releases/2014/05/snapchat-settles-ftc-charges-promises-disappearing-messages-were. It is worth mentioning that the app, except for the photos, retains all (numerical) information about the communication that is taking place: to whom you send how many photos, who takes a screenshot of which image, and who has looked at the photos on My Story. A feature such as Snapstreak—a streak is created when friends share photos over three consecutive days but is destroyed if a day is missed—is an additional, artificial, game-like incentive to increase the use of Snapchat and hence the posting of pictures.

28. "Competing for the same territory of human culture, each claims an exclusive right to make meaning out of the world": narration as "cause-and-effect trajectory of seemingly unordered items (events)"; the database as representation of the world, "as a list of items . . . it refuses to

order," and a "new way to structure our experience of ourselves and of the world." Lev Manovich, *The Language of New Media* (Cambridge, MA: MIT Press, 2002), 225. The paradigm shift indicated here already basically occurs when information is managed on the internet or personal computer, as soon as the filing of documents or access to them no longer occurs taxonomically, by means of a system of files and subfiles, but instead by keywords, via a search engine, through which the necessary inclusion and attribution of elements within a larger whole, which is also characteristic of the narrative model, is lost.

29. Tarleton Gillespie, "The Relevance of Algorithms," in *Media Technologies: Essays on Communication, Materiality, and Society*, ed. T. Gillespie, P. J. Boczkowski, and K. A. Foot (Cambridge, MA: MIT Press, 2014), 171.

30. Manovich, *The Language of New Media*, 219. On the preference for photos, videos, and likes over text updates, see Jean-Sebastien B. Miousse, "How to Get Control on Facebook and How the Algorithms Work," *Science 2.0*, October 19, 2010, http://www.science20.com/science_and _music_your_ears/blog/how_get_control_facebook_and_how_algori thms_work.

31. The "nanopublication" (as a variant of the Semantic Web promoted by Berners-Lee) aims at a computer-friendly, quasi-numerical formalization and classification of statements. See the discussion of this question in my book *Data Love: The Seduction and Betrayal of Digital Technologies* (New York: Columbia University Press, 2016), 82–86.

32. Michael Moorstedt, "Erscanne dich selbst!" [Scan yourself!], in *Big Data. Das neue Versprechen der Allwissenheit* [Big data: The new promise of universal knowledge], ed. Heinrich Geiselberger and Tobias Moorstedt (Frankfurt: Suhrkamp, 2013), 71.

33. Wilhelm Schmid, "Fitness? Wellness? Gesundheit als Lebenskunst" [Fitness? Wellness? Health as the art of living], in *Globalisierung im Alltag* [Everyday globalization], ed. Peter Kemper and Ulrich Sonnenschein (Frankfurt: Suhrkamp, 2002), 214. Michel Foucault, *The Care of the Self*, trans. Robert Hurley (New York: Random House, 1986).

34. The reference to the inscription on the Temple of Apollo in Delphi goes back to the founding father of the Quantified Self movement: Gary Wolf, "Know Thyself: Tracking Every Facet of Life, from Sleep to Mood to Pain, 24/7/365," *Wired*, July 17, 2009; Nicholas Felton, "Numerical Narratives," lecture at the Department of Design, Media,

Arts, UCLA, November 15, 2011, http://video.dma.ucla.edu/video/
nicholas-felton-numerical-narratives/387. The concept "numerical narratives" was previously used in the context of bureaucratically organized information in health care. See Lester Coutinho, Suman Bisht,
and Gauri Raje, "Numerical Narratives and Documentary Practices:
Vaccines, Targets and Reports of Immunisation Programme," *Economic and Political Weekly* 35, no. 8/9 (February 19–26, 2000): 656–66.
Felton, who also developed the above-mentioned diary app "Reporter,"
attracted the attention of various curators thanks to his statistical
reporting on his life. He was included, for example, in the 2011 exhibition "Talk to Me," at the New York Museum of Modern Art, and the
exhibition "Virtual Identities," at the Palazzo Strozzi in Florence. He
also attracted the attention of Facebook, which ultimately hired him to
design the Timeline.

35. Nora Young, *The Virtual Self: How Our Digital Lives Are Altering the
World Around Us* (Toronto: McClelland & Stewart, 2012), 87–88. The
advertising for the athletic shoe Nike+ with a built-in pace tracker
conveys a similar sense of self-encounter: "See all your activity in
rich graphs and charts. Spot trends, get insights and discover things
about yourself you never knew before." http://nikeplus.nike.com/plus
/what_is_fuel. For a different option of consolidation—the "mastery"
of an individual's "drift" in social dynamics through reintegration in
narrative contexts—see Richard Sennett, *The Corrosion of Character:
The Personal Consequences of Work in the New Capitalism* (New York:
Norton, 1998).

36. In light of all this, it seems premature (and rather due to the easily
available pun) to characterize the "dataism" of the measurement model
"digital Dadaism" —as "nihilism" —because, like Dadaism, it gets by
without meaning, as Han suggests when he writes, "Data and numbers are additive, not narrative. Meaning, by contrast, relies on narration." Byung-Chul Han, *Psychopolitics: Neoliberalism and New Technologies of Power*, trans. Erik Butler (New York: Verso Futures, 2017),
52. While the Dadaists responded to the "bankruptcy of ideas" and
erosion of language during the First World War with a refusal of
sense—for this explanation of Dadaist nonsense texts, see the entries
of June 12 and 24, 1916, in Hugo Ball's diary *Flight Out of Time* (Hugo
Ball, *Flight Out of Time: A Dada Diary*, ed. John Elderfield, trans. Ann
Raimes [Berkeley: University of California Press, 1996])—dataism, in
an epistemologically comparable crisis situation, seeks meaning

precisely in the numbers. This may be problematic from the vantage point of philosophy and narrative psychology, but it signals the opposite of nihilism, as the facticity of data responds to the crisis of narration. Deleuze, with an argument resembling that of the erstwhile Dadaists ("Maybe speech and communication have been corrupted. They're thoroughly penetrated by money . . ."), calls for "creating vacuoles of non-communication, circuit breakers, so we can elude control." Gilles Deleuze, "Control and Becoming," in *Negotiations 1972– 1990*, trans. Martin Joughin (New York: Columbia University Press, 1997), 175. On the concept of dataism, see Steve Lohr, *Data-ism: The Revolution Transforming Decision Making, Consumer Behavior, and Almost Everything Else* (New York: Harper Business, 2015).

37. Jill Walker-Rettberg, *Seeing Ourselves Through Technology: How We Use Selfies, Blogs, and Wearable Devices to See and Shape Ourselves* (New York: Palgrave Pivot, 2014), 81, including comments on the "shift from human-generated to machine-generated self-representations" (76).

38. Byung-Chul Han, *The Scent of Time: A Philosophical Essay on the Art of Lingering*, trans. Daniel Steuer (New York: Polity, 2017), 50, 51. On Gatterer's remark, see note 10, above.

39. Han, *Scent of Time*, 51, 53, 54, 11.

40. Han, *Scent of Time*, 17. Han reminds us that the subject of the Enlightenment and modernity was "a free human being that projects itself toward the future. Time is not fate but projection" (15). Han's perspective can be found earlier in Thomas Hylland Eriksen's *Tyranny of the Moment: Fast and Slow Time in the Information Age* (London: Pluto, 2001), when Eriksen describes the "loss of time" (47) as a "Lego brick syndrome" (121). On living on toward "the fullnes [*sic*] of the 'years of the Lord,'" see White, *The Content of Form*, 11. On the end of history from the perspective of the philosophy of history (according to which political events no longer change the basic structure of society), see Francis Fukuyama, *The End of History and the Last Man* (New York: Avon, 1993). Liu compares the entries in the annals that are cited by White with Twitter posts, as a form of factually oriented, non-narrative history writing: "*Now* is the order of the day. *Now* is history as it really *is*, with no *was* in view more extensive than—on a typical Web 2.0 screen—just a handful of entries ordered by most-recent at top." Alan Liu, "Friending the Past: The Sense of History and Social Computing," *New Literary History* 42, no. 1 (2011): 20.

41. Han, *Scent of Time*, 46, VII.

42. Siegfried Kracauer, "Those Who Wait," in *Mass Ornament: Weimar Essays*, trans. Thomas Y. Levin (Cambridge, MA: Harvard University Press, 2005), 129. See also "Cult of Distraction: On Berlin's Picture Palaces," in the same collection, 323–28. The following quotes are drawn from "Those Who Wait," 135–38. On metaphysical homelessness, see the section "Shelter for the Homeless," in Kracauer's *The Salaried Masses: Duty and Distraction in Weimar Germany*, trans. Quintin Hoar (London: Verso, 1998), 88–95.

43. Siegfried Kracauer, "Boredom," in *Mass Ornament*, 332, 334 (translation modified).

44. Friedrich Nietzsche, *Thus Spake Zarathustra*, trans. Thomas Common (New York: Modern Library, n.d.), 46. Nietzsche's encouragement of laziness is understandable in the context of various versions of the "praise of leisure," which appeared both in early Romanticism, for example as the name of a chapter in Friedrich Schlegel's novel *Lucinda* (1799); and in heretical Marxism: Paul Lafargue, Karl Marx's son-in-law, titled an essay "The Right to Be Lazy." Also of interest in this context is Walter Benjamin's praise of boredom in his essay "The Storyteller": "Boredom is the dream bird that hatches the egg of experience." Walter Benjamin, "The Storyteller: Observations on the Works of Nikolai Leskov," in *Selected Writings*, vol. 3: *1935–1938*, ed. Howard Eiland and Michael W. Jennings (Cambridge, MA: Harvard University Press, 2002), 149. Wittkower, in his essay "Boredom on Facebook," starts by imagining boredom, with Schopenhauer, Nietzsche, and Heidegger, as an existential threat: "Being bored by something motivates a break, a change. . . . We shy away from this existential boredom." Dylan E. Wittkower, "Boredom on Facebook," in *Unlike Us Reader: Social Media Monopolies and Their Alternatives*, ed. Geert Lovink and Miriam Rasch (Amsterdam: Institute of Network Cultures, 2013), 183. But he understands Facebook—unlike the argument of this volume—not as an antidote but as the site of boredom: "One of the great successes of Facebook is the way in which it allows us to be bored together" (185). This counterintuitive conclusion is based on a sudden, unclear shift in argumentation from the concept of boredom as a lack of motivation and distraction to the notion of non-goal-directed activity, when "boredom" is suddenly welcomed as "'hanging out' and 'quality time'" (184), "friendertainment" (185), or "leisure well but purposelessly spent" (187). With this interpretation of leisure, Wittkower properly celebrates Facebook (like Pschera's apology on

behalf of social media) as a means of nonintentional, phatic communication but blocks the insight toward which Kracauer's and Nietzsche's concepts of boredom are directed: that Facebook is anything but the place where a person "comes to himself."

45. Rosa, *Weltbeziehungen*, 224, 218. Rosa defines situational identity as the "self-understanding corresponding to the temporalized time of late modernity," with which a person's own life "is no longer experienced as a progressively unfolding (and plannable) project, but as open 'play,' or 'drift,' in which all identity predicates require a temporal index—'at the moment, you are married to X'; or 'are,' at the moment, a graphic artist; or, at the time, voted for the Greens, etc."

46. Zygmunt Bauman: "From Pilgrim to Tourist—or a Short History of Identity," in *Questions of Cultural Identity*, ed. Stuart Hall and Paul du Gay (London: Sage, 1996), 25.

47. Galen Strawson, "Against Narrativity," in *The Self?*, ed. Galen Strawson (Malden, MA: Blackwell, 2005), 65–86. See also Strawson's concept of the "thin subject" in his book *Selves: An Essay in Revisionary Metaphysics* (Oxford: Oxford University Press, 2009). An attractive addressee for Strawson's intervention is the British moral philosopher Alasdair MacIntyre, who conceived of the "unity of a human life" as "unity of a narrative quest" ("Against Narrativity," 71). Strawson also makes reference to Jerry Bruner, Marya Schechtmann, Paul Ricœur ("How, indeed, could a subject of action give an ethical character to his or her own life taken as a whole if this life were not gathered together in some way, and how could this occur if not, precisely, in the form of a narrative?" 71), and Charles Taylor ("a basic condition of making sense of ourselves is that we grasp our lives as a *narrative*," 70).

48. Strawson, "Against Narrativity," 67. With the turn against Heidegger, Strawson moves the discussion back to its actual source, for the perspective of an internally structured consciousness based on the unity of past, present, and future is the foundation of phenomenology, from which Ricœur develops his philosophy of narrativity. Strawson, on the contrary, is closer to David Hume's "bundle theory," according to which the self is no more than a series of lived experiences. (Strawson published his research on this topic, which was conducted at approximately the same time as his essay, in 2011 under the title *The Evident Connexion: Hume on Personal Identity* (Oxford: Oxford University Press, 2011). Kant had already opposed this

association theory of consciousness by positing the synthesis of experiences in a unity of consciousness. Unsurprisingly, Strawson's antinarrative stance has been greeted with skepticism and rejection by narrative psychologists. He received support for his doubt that a life experienced as coherent is in itself already ethical, arguing that National Socialism, among other things, decisively proved that narrative meaningfulness can be present without regard for ethical values and critical reflection. See Hanna Meretoja, "Narrative and Human Experience. Ontology, Epistemology, and Ethics," *New Literary History* 45, no. 1 (Winter 2014), 102–103. Meretoja launches a serious objection to Strawson's dichotomous opposition between diachronic and episodic types when she argues that individuals do not organize their experiences (or historical facts) narratively in retrospect, as Strawson's epistemological definition of the narrative assumes, but already experience things narratively, according to the ontological definition of narrativity (96).

49. Philippe Lejeune, "Autobiography and New Communication Tools," in *Identity Technologies: Constructing the Self Online*, ed. Anna Poletti and Julie Rak (Madison: University of Wisconsin Press, 2014), 247–58. Lejeune's analysis of the present is reminiscent of Bauman's: "We are losing our long-term connections, our rootedness in the past, and the ability to project ourselves into the future, all of which allowed us to construct a narrative identity. We are skating along swiftly in a present that annihilates the past and denies the future" (250).

50. On the "authority of the form," see also Reichert, *Macht der Vielen*, 61. McNeill characterizes the algorithms on Facebook as "shadow biographers, telling users about themselves while telling the site and its advertisers about the users." "Agency," she concludes, "seen as so key to the humanist subject, has been transferred to the software that reads and produces users. Where, indeed, do we end and Facebook begin?" Laurie McNeill, "There Is No 'I' in Network: Social Networking Sites and Posthuman Auto/Biography," *Biography* 35, no. 1 (Winter 2012), 75, 79. The network friends also participate in writing one's own Facebook biography to the extent that their likes lend weight to updates and in the timeline, for example, can bring them back to the present. The finding of posthuman, postactive writing on social networks speaks against the assumption that Facebook's timeline interface might once again make narration, in spite of or perhaps in cooperation with the database, into a contemporary form of world

and self-understanding. Also compare van Dijck ("You Have One Identity," 204–207), where she observes that the chronological order of the entries and the possibility to delete selected postings hardly constitute narrative form and control, which also explains why employers are more interested in the Facebook accounts than the LinkedIn records of potential employees (212).

51. "Perhaps personal narrative, then, to borrow Katherine Hayles's description of humans, 'has always been posthuman' (291), a prospect that makes the apparently paradoxical a productive frame for rethinking how we craft and consume selves." McNeill, "There Is No 'I' in Network," 80. McNeill cites N. Katherine Hayles's description of the relinquishment of human agency, in "distributed cognitive systems," to "nonhuman actors" and draws on Lejeune's concept of an autobiographical pact to propose the concept of a "posthuman pact" for Facebook's "algorithmic autobiographies" (80, 75). The term "blackboxing" is drawn from an essay by Galloway that contains the phrase "blackboxing of the self." Alexander R. Galloway, "Black Box, Black Bloc," in *Communication and Its Discontents: Contestation, Critique, and Contemporary Struggles*, ed. Benjamin Noys (London: Minor Compositions, 2011), 237–52. The German translation of Galloway's phrase ("'Blackboxing' des Selbst") is even more suggestive, since it can be read to refer to the self as both subject and object of blackboxing—in the system of digital data streams, the self is subject to the creation of a profile that it neither knows nor controls, while at the same time it consciously submits to the software as an external autobiographer. Alexander R. Galloway, "Black Box, Schwarzer Bloc," in *Die technologische Bedingung. Beiträge zur Beschreibung der technischen Welt* [The technological condition: Contributions to the description of the technical world], ed. Erich Hörl (Berlin: Suhrkamp, 2011), 273.

52. Roland Barthes, *Writing Degree Zero*, trans. Annette Lavers and Colin Smith (New York: Hill and Wang, 1968), 31, 32. On the reduction of reality by means of narrative ordering in literature as well as historiography, see the chapter "Writing and the Novel" (29–40). Barthes's own "postmodern autobiography" (as Hayden White termed it), with its fragmentary and fictional tendencies, narrated in the third person and interspersed with various aspects and associations that have no evident connection, like a hypertext, follows from his critique. On "actual, inner" truth in Wilhelm von Humboldt, see "On the Historian's Task [1821]."

53. Julia Kristeva, *Strangers to Ourselves*, trans. Leon S. Roudiez (New York: Columbia University Press, 1991), 182: "How could one tolerate a foreigner if one did not know one was a stranger to oneself?" The argument about greater objectivity was foreshadowed by Hayles's discussion of the greater reliability of "distributed cognition" as compared to purely subjective perception and problem solving. N. Katherine Hayles, *How We Became Posthuman: Virtual Bodies in Cybernetics, Literature, and Informatics* (Chicago: University of Chicago Press, 1999), 288ff. Hayles's argumentation refers to and contests Joseph Weizenbaum's warning that machines could take control at some point: "The prospect of humans working in partnership with intelligent machines is not so much a usurpation of human right and responsibility as it is a further development in the construction of distributed cognition environments, a construction that has been ongoing for thousands of years" (289–90). Hayles further develops the notion of "distributed cognition" in her more recent essay on the "nonconscious cognition of intelligent devices," as exemplified by the "smart house" and "self-driving car" but also by "evolutionary algorithms." N. Katherine Hayles, "Cognition Everywhere: The Rise of the Cognitive Nonconscious and the Costs of Consciousness," *New Literary History* 45, no. 2 (2014): 211, 202. Along with the advantages of the cognitive unconscious—"nonconscious cognition operates without the biases inherent in consciousness" (214)—Hayles also notes the risk of unconscious conditioning in the context of "affective capitalism" (212) but essentially remains more positively inclined than the representatives of the critical theory of digital media, who consider the externalization of self-representation to be loss of agency: "Elements of subjectivity, judgment and cognitive capacities are increasingly delegated to algorithms and prescribed to us through our devices, and there is clearly the danger of a lack of critical reflexivity or even critical thought in this new subject." David M. Berry, *Critical Theory and the Digital* (New York: Bloomsbury, 2014), 11. Chapter 3 will explain why this loss of reflexivity should be understood more as a danger to social cohesiveness than as the overcoming of individual inadequacies. The epilogue will come back to Hayles's perspective as a possible mode of operation in the future.

54. Douglas Coupland, *Generation X: Tales for an Accelerated Culture* (New York: St. Martin's, 1991), 8.

3. DIGITAL NATION

1. Siegfried Kracauer, "Photography," trans. Thomas Y. Levin, *Critical Inquiry* 19, no. 3 (Spring 1993): 432, 433, 427 (translation modified).

2. Jean Baudrillard, *Photographies 1985–1998* [exhibition catalog], ed. Peter Weibel, trans. Susanne Baumann et al. (Ostfildern-Ruit: Hatje Cantz, 1999), 132.

3. Kracauer, "Photography," 433; Baudrillard, *Photographies*, 25. The self-presentation of things is (merely) "apparent" in two senses. For one thing (this is an old debate), photographs are never wholly indexical, as Peirce (before Kracauer), or "uncoded," as Barthes (after Kracauer), pointed out. For another, the cultural denotation and connotation of the photographs is evident precisely in illustrated magazines, as the frame within which the objects are displayed. Kracauer's critique rests on an epistemological remark on the historical progression of consciousness "from substance and matter to the spiritual and the intellectual" (433), in other words, from entanglement in nature to conceptual, abstract thought. This process of emancipation is halted and reversed by photography, with which the "foundation of nature devoid of meaning" (434) regains ground, as the cognition encouraged by conceptual consciousness gets lost in the material evidence of the images. Later, Adorno and Horkheimer will discuss the "demythologization" and "rationalization" of language as a falling silent, holding that "the more words change from substantial carriers of meaning to signs devoid of qualities . . . the more purely and transparently they designate what they communicate" and that "the blindness and muteness of the data to which positivism reduces the world passes over to language itself, which is limited to registering those data." Theodor W. Adorno and Max Horkheimer, *Dialectic of Enlightenment*, ed. Gunzelin Schmid Noerr, trans. Edmund Jephcott (Stanford, CA: Stanford University Press, 2002), 133–34. The "foundation of nature devoid of meaning" of photography and the "muteness" of positivism return in the model of data objectivity.

4. Bernard Stiegler, *Taking Care of Youth and the Generations*, trans. Stephen Barker (Stanford, CA: Stanford University Press, 2010). Stiegler's argument is that only *deep* attention, with its specific form of synapse generation, allows the transition to maturity. For the allusion to Arendt and Eichmann in the following sentence, see Christian

Lotz, "Review of Bernard Stiegler, *The Re-Enchantment of the World: The Value of Spirit Against Industrial Populism*, trans. Trevor Arthur (London: Bloomsburg Academic, 2014)," *Marx & Philosophy: Review of Books 2015* (March 11, 2015), http://marxandphilosophy.org.uk/review ofbooks/reviews/2015/1754.

5. Bernard Stiegler, *For a New Critique of Political Economy*, trans. Daniel Ross (Cambridge: Polity, 2010), 5; Bernard Stiegler, *The Re-Enchantment of the World: The Value of Spirit Against Industrial Populism*, trans. Trevor Arthur (London: Bloomsbury Academic, 2014), 18. See also Bernard Stiegler, "Care," in *Telemorphosis: Theory in the Era of Climate Change*, ed. Thomas Cohen (Ann Arbor, MI: Open Humanities Press, 2012), 1:104–20.

6. For a discussion of the "linguistic milieu" of "participation," see Bernard Stiegler, "Memory," in *Critical Terms for Media Studies*, ed. W. J. T. Mitchell and Mark B. N. Hansen (Chicago: University of Chicago Press, 2010), 83–84. On the concept of "cognitive and affective proletarianization," see Stiegler, *For a New Critique of Political Economy*, 30; on "technical memory" as "generalized proletarianization induced by the spread of hypomnesic technologies," see 35. Kracauer's concept of "spatial appearance" can also be applied, in a certain sense, to the presence of Facebook friends on a person's own Facebook page. No longer do I describe my friends to third persons (although this also occurs, in the form of commentaries and likes); instead, the friends present themselves to these third persons as their posts become available to them on my page.

7. Nicholas Carr, *The Shallows: What the Internet Is Doing to Our Brains* (New York: Norton, 2011), 137, 122.

8. The immediate example of this perspective is Steven Johnson, *Everything Bad Is Good for You: How Today's Popular Culture Is Actually Making Us Smarter* (New York: Riverhead, 2005). But the real model is Walter Benjamin's 1936 essay "The Work of Art in the Era of its Technological Reproducibility," in which Benjamin responded to his contemporaries' common critique that cinema would destroy contemplation by arguing that the "intensified presence of mind" with which audiences responded to the "shock effect of the images" represented the very exercise of the apperceptive faculty that was required by the accelerated pace of life in an early-twentieth-century metropolis. Those who are familiar with Benjamin's other writings will know that

he was defending the new medium not only against its conservative critics but also against his own conviction.

9. Stiegler, "Memory," 78. On the hope of "associated hypomnesic milieus of digital networks . . . insofar as they are cooperative and participative," see 84. In this context, one should be wary of overly dichotomous views of the opposition sender : receiver or consumer : producer. Being a sender does not yet signify reflexivity (or conscious coding); the unending flow of status updates on the social network often differs from the unreflexive flow of images from the culture industry only when it is viewed from an actionistic perspective. The trick of the culture industry in the age of "communicative capitalism" is the continuation of distraction (from thinking) by means of interaction. The "We Media" of the Web 2.0 do not, in themselves, necessarily strengthen the opposition of an oppressed community to a social order that is being criticized. Often, instead, they may lead to a drowning of critique in the busyness of banal sociality.

10. Stanley Aronowitz, "Looking Out: The Impact of Computers on the Lives of Professionals," in *Literacy Online: The Promise (and Peril) of Reading and Writing with Computers*, ed. Myron C. Tuman (Pittsburgh, PA: University of Pittsburgh Press, 1992), 133. Bolter saw hypertext as "a vindication of postmodern literary theory." Jay David Bolter, "Literature in the Electronic Writing Space," in *Literacy Online: The Promise (and Peril) of Reading and Writing with Computers*, ed. Myron C. Tuman, (Pittsburgh, PA: University of Pittsburgh Press, 1992), 24; Landow (whose influential monograph *Hypertext 2.0* already made the connection in its subtitle) saw the new technology as embodying the ideas of Barthes, Derrida, and Foucault. George P. Landow, *Hypertext 2.0: The Convergence of Contemporary Critical Theory and Technology* (Baltimore, MD: Parallax, 1997), 91. For an extensive critique see my study *Interfictions. Vom Schreiben im Netz* [Interfictions: On writing on the net] (Frankfurt: Suhrkamp, 2002).

11. Isaiah Berlin, *The Hedgehog and the Fox: An Essay on Tolstoy's View of History*, trans. Henry Hardy (Princeton, NJ: Princeton University Press, 2013), 115–16; Ben Macintyre, "We Need a Dugout Canoe to Navigate the Net," *Times* (London), January 28, 2010. The icon on the Firefox browser is naturally not a reference to Berlin but a stopgap, since the name that was originally chosen, Firebird, was already spoken for. Browsers, which, following Macintyre, could be seen as

technical realizations of the fox, are actually ambivalent in their relation to Berlin's distinction. The bird could serve as a symbol of openness and exploratory pleasure, while the compass of the Safari browser,
with its orientation to a specific goal, is more reminiscent of the
hedgehog.

12. The writer Jean Paul, whose real name was Jean Paul Friedrich Rich-
ter, was a prolific and eccentric writer and a convinced democrat. He
became one of the most popular authors of his era, and his works are
known for their witticisms and ironic characters.—Trans.

13. Georg Wilhelm Friedrich Hegel, *Aesthetics: Lectures on Fine Art*,
2 vols., trans. T. M. Knox (Oxford: Clarendon, 1975), 295.

14. Jean Paul, "Clavis Fichtiana," in *Werke*, part 1, vol. 3, ed. Norbert
Miller (Munich: Hanser, 1963), 1,028; Jean Paul, "Briefe und bevorstehender Lebenslauf" [Letters and forthcoming biography], in *Werke*,
part 2, vol. 4, ed. Norbert Miller (Munich: Hanser, 1963), 1,022.

15. In 1780, when he was only seventeen, under the title "Jeder Mensch ist
sich selbst Masstab, wonach er alles äussere abmist" [Every human
being is his own measure, by which he measures everything external],
Jean Paul already described individual systems of thought and con-
ceptuality not only as different from one another but as incompatible
and untranslatable.

16. Jean Paul, *Levana; Or, the Doctrine of Education*, trans. A. H. (London: George Bell and Sons, 1891).

17. The reference to wit as a priest and the soul of wit is drawn from Jean
Paul, "Vorschule der Aesthetik" [Introduction to aesthetics], in
Werke, vol. 5, ed. Norbert Miller (Munich: Hanser, 1963), 173. Jean
Paul calls for the "development of wit" in children because wit "by the
pleasure of discovery . . . gives increased power of command over . . .
ideas," while normally "children are taught more ideas than command
over those ideas" (*Levana*, 382–83). The following quotations on wit
are found on pages 204 und 205.

18. Jean Paul, *Levana*, 550. I discuss Jean Paul's cosmopolitan information model in the context of Herder and Fichte in my essay "System
und Witz—Jean Pauls Kosmopolitismus als Effekt des sprachphilosophischen Zweifels" [System and wit: Jean Paul's cosmopolitanism
as an effect of linguistic-philosophical doubt], in *Kulturelle Grenzziehungen im Spiegel der Literaturen: Nationalimus, Regionalismus, Fundamentalismus* [Drawing cultural boundaries in the mirror of literatures:
Nationalism, regionalism, fundamentalism], ed. Horst Turk, Brigitte

Schultze, and Roberto Simanowski (Göttingen: Wallstein, 1999), 170–92; and, with reference to new media, in my essay "Jean Pauls vergebliche Postmodernität" [Jean Paul's futile postmodernity], in *Jahrbuch der Jean Paul Gesellschaft* [Yearbook of the Jean Paul Society] (2013), 61–73, from which the current line of argument is drawn.

19. On the concept of cosmopedia, see Pierre Lévy, *Collective Intelligence: Mankind's Emerging World in Cyberspace*, trans. Robert Bononno (Cambridge, MA: Perseus, 1997). The Jean Paul obituary is from Ludwig Börne, "Denkrede auf Jean Paul" [Commemorative address for Jean Paul], ed. Karl Rauch (Bern: Franke, 1964), 6.

20. Arthur Schopenhauer, "On Thinking for One's Self," in *Essays of Arthur Schopenhauer*, ed. and trans. T. Bailey Saunders (New York: A. L. Burt, 1902), 325.

21. Johann Gottfried Herder, "Briefe zu Beförderung der Humanität" [Letters for the advancement of Humanity], in *Herders Sämmtliche Werke*, ed. Bernhard Suphan (Berlin: Weidmann, 1883), 18:90–91.

22. Jean Paul, *Levana*, 100 (fragment 1, chap. 16).

23. Schopenhauer, "Thinking for One's Self," 323. As insightful as Schopenhauer's warning may sound, in an era when knowledge is mainly gleaned from significantly shorter texts, it is already a lot if an individual actually musters the necessary intellectual effort and cognitive stamina to approach the mental cosmos of an entire book.

24. Gianni Vattimo, *The Transparent Society*, trans. David Webb (Baltimore, MD: Johns Hopkins University Press, 1992), 5, 6. The biblical metaphor Herder and Schopenhauer used to describe the confusion created by excessive media availability retained its power in descriptions of the "paradisiacal" beginnings of radio, when, for example, radio is portrayed, in a period of "Babylonian confusion," as a "gigantic megaphone" bringing the "multiplicity, the mixture of voices and calls under its single, wave-saturated spell." Comment of the director of the Silesian Radio, Fritz Walter Bischoff, in 1929, cited by Albert Kümmel, "Innere Stimmen. Die deutsche Radiodebatte" [Inner voices: The German radio debate], in *Einführung in die Geschichte der Medien* [Introduction to media history], ed. Albert Kümmel, Leander Scholz, and Eckhard Schuhmacher (Paderborn: UTB, 2004), 176–77.

25. Vattimo, *The Transparent Society*, 7 (translation modified).

26. Jean Paul, "Vorschule der Aesthetik," 206. In the preface to his *Biographische Belustigungen* [Biographical amusements] (1795), in *Werke*,

part 2, vol. 4, ed. Norbert Miller (Munich: Hanser, 1963), 355, Jean Paul contrasts attentive reading with the fleeting gaze he ascribes to "girls." In the text, refractory readers hail Jean Paul before an imaginary court of law to compel him to stick to the matter at hand and in his future writing to avoid such excessive digressions. The following quotation is from Jean Paul, "Die Taschenbibliothek" [The pocket library], in *Werke*, part 2, vol. 3, ed. Norbert Miller (Munich: Hanser, 1963), 772.

27. The concept of collective memory refers to the memory of a group of people. Like individual memory, it is formed from communicative memory (person-dependent, biographical, primarily orally transmitted recollections that do not go back further than three generations) and cultural memory (written, visual, and since the twentieth century also electronically stored cultural practices and mediated references to the past). The concept of cultural memory (*kulturelles Gedächtnis*) that is used in the following generally refers to groups; the concept of cultural recollection (*kulturelle Erinnerung*) usually refers to cultural memory. On the distinction between storage memory (as the collection of all transmitted materials) and functional memory (the dominant cultural memory), see Aleida Assmann, *Memory and Political Change*, trans. Linda Shortt (New York: Palgrave Macmillan, 2012).

28. Wolfgang Ernst, "Das Archiv als Gedächtnisort" [The archive as a site of memory], in *Archivologie. Theorien des Archivs in Theorie, Medien und Künsten* [Archivology: Theories of the archive in theory, media and the arts], ed. Knut Ebeling and Stephan Günzel (Berlin: Kadmos, 2009), 168. Pierre Nora's concept of the "lieu de mémoire," where collective memory takes place and/or is transmitted, refers not only to places (geographical reference points, monuments, museums) but also to times (anniversaries), rituals, artworks, persons, and narratives.

29. Kracauer, "Photography," 61. Friedrich Nietzsche, "On the Use and Abuse of History for Life," in *Untimely Meditations*, trans. R. J. Hollingdale (Cambridge: Cambridge University Press, 1997), part 4, para. 3.

30. On the concept of exosocialization as the site of "production and reproduction of men outside local intimate units," see Ernest Gellner, *Nations and Nationalism* (Ithaca, NY: Cornell University Press, 1983), 38; on the biography of the nation, see Benedict Anderson, *Imagined Communities: Reflections on the Origin and Spread of Nationalism* (New York: Verso, 1996).

31. Kracauer, "Photography," 61.

32. "The default value that is automatically realized if a person does not opt for something else and expends neither energy nor attention is now remembering, not forgetting." Elena Esposito, "Die Formen des Web-Gedächtnisses. Medien und soziales Gedächtnis" [The forms of Web-memory: Media and social memory], in *Formen und Funktionen sozialen Erinnerns. Sozial- und kulturwissenschaftliche Analysen* [Forms and functions of social remembering: Social and cultural studies analyses], ed. René Lehmann, Florian Öchsner, and Gerd Sebald (Wiesbaden: Springer VS, 2013), 92.

33. On the "obsession with the archive," see Pierre Nora, "Between Memory and History: Les Lieux de Mémoire," *Representations* 26 (Spring 1989): 13; and the issue "The Storage Mania" of the journal *Mediamatic* 8, no. 1 (Summer 1994). On the "explosion of memory discourses" and "culture of memory," see Andreas Huyssen, *Present Past: Urban Palimpsests and the Politics of Memory* (Stanford, CA: University of California Press, 2003), 4, 15. The quotation on the shift "from present futures to present pasts" is found in the same volume, 11. The concept "musealization" (*Musealisierung*) was coined by Lübbe and refers to a form of compensation for the "loss of familiarity" that goes hand in hand with accelerated modernity. Hermann Lübbe, *Im Zug der Zeit: Verkürzter Aufenthalt in der Gegenwart* [In the course of time: Abbreviated stay in the present] (Berlin: Springer, 1992).

34. Nora, "Between Memory and History," 13. Nora distinguishes between history (as the neutral "representation of the past") and memory ("a bond tying us to the eternal present . . . affective and magical") (8). His conclusion corresponds to Postman's regretful conclusion that narrative history has come to an end. Neil Postman, *Technopoly: The Surrender of Culture to Technology* (New York: Vintage, 1992). Postman asks history teachers to become "histories teachers," in other words, to show "how the religion, politics, geography, and economy of a people lead them to re-create their past along certain lines," and he notes, "To teach the past simply as a chronicle of indisputable, fragmented, and concrete events is to replicate the bias of Technopoly, which largely denies our youth access to concepts and theories, and to provide them only with a stream of meaningless events" (191). For the "emphatic site of history" and "passionless archive," see Ernst, "Das Archiv als Gedächtnisort," 168.

35. Friedrich Kittler's 1985 book *Aufschreibesysteme 1800/1900* [Writing systems] was translated into English as *Discourse Networks 1800/1900*,

trans. Michael Metteer with Chris Cullens (Stanford, CA: Stanford University Press, 1990). For Kittler, *Aufschreibesysteme* were "the network of technologies and institutions that allow a given culture to select, store, and process relevant data" (369). For Nora, modern memory is "archival," since it depends on material traces, direct recording, and the visual givens of the image: "What began as writing ends as high fidelity and tape recording. The less memory is experienced from the inside the more it exists only through its exterior scaffolding and outward signs—hence the obsession with the archive that marks our age, attempting at once the complete conservation of the present as well as the total preservation of the past" ("Between Memory and History," 13). The question is whether the new technologies of archiving are a reaction to the lost capacity to remember or its cause. On digital media, archiving is mostly automatic, a result of failure to delete, as personal e-mail archives demonstrate. On "self-musealization per video recorder," see Huyssen, *Present Past*, 14.

36. Wolfgang Ernst, *Das Gesetz des Gedächtnisses. Medien und Archive am Ende (des 20. Jahrhunderts)* [The law of memory: Media and archive at the end (of the twentieth century)] (Berlin: Kulturverlag Kadmos, 2007), 264, 282, 284, 278. Jochum notes that "The total recall that the Internet promised is . . . anything but the implementation, through data technology, of the 'memory of humankind.' Instead, we are looking at a global denial of memory. The place of scholarly communities or the interested publics working on cultural memory is taken by potential technical access to a global data network that offers information, not knowledge." Uwe Jochum, "Die virtuelle Bibliothek" [The virtual library], in *7 Hügel*, vol. 6: *Wissen* [Knowledge] [catalogue of the exhibition] (Berlin, 2000), 40; cited in Ernst, *Das Gesetz des Gedächtnisses*, 277.

37. Elena Esposito, *Soziales Vergessen. Formen und Medien des Gedächtnisses der Gesellschaft* [Social forgetting: Forms and media of the memory of society] (Frankfurt: Suhrkamp, 2002), 358. On "mass customization," see 301. Also: "The static model of data storage is replaced by the dynamic model of data construction, which is gradually created based on the commands of the user: by the model that is realized on the internet by the search engines" (257).

38. For this reason, Esposito (*Soziales Vergessen*, 318, 351) talks about an "autological model of memory," which, via a feedback loop that is

scarcely controllable, makes collective memory conform to the individual's perspective.

39. Ernst, *Das Gesetz des Gedächtnisses*, 264. This situation results, for Ernst, in the need for a new form of ideology critique, which is carried out, inter alia, in software studies, by interrogating the cultural factors in programming. Examples of collective memory that are made possible by the internet include Steven Spielberg's Shoah project (http://sfi.usc.edu), which archives memories of Holocaust survivors in video form; the Vietnam Veterans Memorial *The Virtual Wall*, which recalls veterans of the Vietnam War through individually submitted visual and written material (http://www.virtualwall.org); or the Mukurtu software project for the digital archiving of the cultural heritage of indigenous communities (http://www.mukurtu.org). On the paradox of "long-distance nationalism," see Benedict Anderson, "The New World Disorder," *New Left Review* 193 (1993): 3–13; Benedict Anderson, "Long-Distance Nationalism: World Capitalism and the Rise of Identity Politics," Wertheim Lecture, University of Amsterdam, 1992.

40. Pierre Nora, "Reasons for the Current Upsurge in Memory," *Eurozine*, April 19, 2002, http://www.eurozine.com/reasons-for-the-current-upsurge-in-memory.

41. A note on the concept of culture, which will play a prominent role in the following discussion: Culture can be understood as civilization if it is given a qualifying, temporal turn, in the sense that we are (already) civilized instead of (still) uncultivated and uneducated. Or it can also be quantifiably and spatially qualified, to refer to systems of value and behavioral norms that create identification and demarcation. One conceptual consequence that derives from the latter approach is "multiculturalism," whereas the "culture of forgetting" (or "hybrid culture") tends to consider civilization, in a globalized, postmodern period, as no longer belonging exclusively to any *one* culture. This skepticism toward a *single* (universal) value system will be the subject of the following discussion of cosmopolitanism, which will not examine either the specificity of "new," realpolitik-based cosmopolitanism or the possibly conflicting obligations that can arise even when value systems are not in doubt—for example, the decision whether to save a Jew from the Nazis if it endangers one's own family.

42. Nora, "Reasons for the Current Upsurge in Memory," 7. Startwell offers an extensive critique of narrativity (with the accent on Alasdair MacIntyre's ethic of narrativity): Crispin Startwell, *End of Story: Toward an Annihilation of Language and History* (Albany: State University of New York Press, 2000), where he counterposes loose sequences of events and the haiku ("which is always devoted to bringing the moment home") to narratives (17).

43. Judith Butler, *Giving an Account of Oneself* (New York: Fordham University Press, 2005), 64, 39.

44. Seyla Benhabib, *Claims of Culture: Equality and Diversity in the Global Era* (Princeton, NJ: Princeton University Press, 2002), 7.

45. Nora, "Between Memory and History," 8–9.

46. Gotthold Ephraim Lessing, *Lessing's Masonic Dialogues (Ernst und Falk)*, trans. Abraham Cohen (London: Baskerville, 1927), 46–47.

47. Johann Gottlob Fichte, *Addresses to the German Nation*, trans. R. F. Jones and G. H. Turnbull (Chicago: Open Court, 1922), 37.

48. Daniel Levy and Nathan Sznaider, *Human Rights and Memory* (Princeton, NJ: Princeton University Press, 2010), 1–2. See also Samuel Moyn, *The Last Utopia: Human Rights in History* (Cambridge, MA: Harvard University Press, 2010); and Julien Benda, *The Treason of the Intellectuals*, trans. Richard Aldington (New Brunswick, NJ: Transaction, 2007), which, in 1927, called for humankind's ascent from its national origins to a European and finally a universal identity (as an overcoming of historical-cultural contexts) and saw the betrayal of the intellectuals in their desire to be an abode for specific (cultural, political, religious) groupings, rather than for all.

49. Charles Taylor, "The Politics of Recognition," in *Multiculturalism: Examining the Politics of Recognition*, ed. Charles Taylor and Amy Gutmann (Princeton, NJ: Princeton University Press, 1994), 44, 62. That Western individualism is not able to claim universality from another (Confucian) perspective was demonstrated by the discussion about "Asian values" in the 1990s and the "Bangkok Declaration" of 1993, as a relativization of the UN Charter of Human Rights.

50. In this sense, Levy and Sznaider expressly oppose postmodern deconstruction and support the continuation of the modernist project (*Human Rights and Memory*, 7), and Deleuze warns, "in philosophy we're coming back to eternal values, to the idea of the intellectual as custodian of eternal values.... These days it's the rights of man that provide our eternal values. It's the constitutional state and other notions

everyone recognizes as very abstract. And it's in the name of all this that thinking's fettered, that any analysis in terms of movements is blocked." Gilles Deleuze, "Mediators," in *Negotiations 1972–1990*, trans. Martin Joughin (New York: Columbia University Press, 1997), 121–22.

51. Butler, *Giving an Account of Oneself*, 79.

52. Gotthold Ephrahim Lessing, *Nathan the Wise*, trans. Adolphus Reich (London: A. W. Bennett, 1860), 112.

53. Christoph Türcke, "Die geheime Kraft des Rings" [The secret power of the ring], in *Lessing. Nachruf auf einen Aufklärer. Sein Bild in der Presse der Jahre 1781, 1881 und 1981* [Lessing: Obituary of an Enlightenment man: His image in the press in years 1781, 1881, and 1981], ed. Klaus Bohnen (Munich: Wilhelm Fink, 1982), 155. The following quotation is from the same source.

54. Norbert Bolz, *Das konsumistische Manifest* [The consumerist manifesto] (Munich: Wilhelm Fink, 2002), 14 ("pragmatic cosmopolitanism"), 16 ("immune system of world society"). The *Nathan* production referenced here was directed by Claus Peymann at the Berliner Ensemble in 2003. The comparison of tolerance and ideological pluralism (as "nonbinding commitment") with the model of capitalist consumption has a tradition that reaches back well beyond Bolz und Türcke. See, for example, Christopher Lasch, *The Minimal Self: Psychic Survival in Troubled Times* (New York: Norton, 1984), 38: "The pluralist conception of freedom rests on the same protean sense of the self that finds popular expression in such panaceas as 'open marriage' and 'nonbinding commitments.' Both originate in the culture of consumption."

55. Lessing, *Nathan the Wise*, 202.

56. The contemporary critique of the concept of cosmopolitanism is a variation on Dajah's opposition to Nathan's "illoyal loyalty" to all humans, using the same descriptors and the example of Nathan, specifically in regard to Dajah, to question the inclusivity of his tolerance model. The critics argue that Nathan oppresses Dajah just as stubbornly as the Patriarch ignores Nathan, because Dajah cannot be integrated into his model of the religion of reason. See Wilfried Wilms, "The Universalist Spirit of Conflict: Lessing's Political Enlightenment," *Monatshefte* 94, no. 3 (2002): 309–10. On the pros and cons of cosmopolitanism, with an essay by Martha C. Nussbaum serving as an example, see Joshua Cohen, ed., *For Love of Country* (Boston: Beacon, 1996).

57. On "cosmopolitanism from below," based on the example of Mumbai, see Arjun Appadurai, *The Future as Cultural Fact: Essays on the Global Condition* (London: Verso, 2013), 198; and Homi K. Bhaba, "Unsatisfied: Notes on Vernacular Cosmopolitanism," in *Text and Nation*, ed. Laura Garcia-Morena and Peter C. Pfeiffer (London: Camden House, 1996), 191–207. On "cosmopolitanization" as "actually existing," "banal" cosmopolitanism versus cosmopolitanism as a normative theory (of the Enlightenment, of intellectuals and politicians), see Ulrich Beck, *Cosmopolitan Vision*, trans. Cioran Cronin (Cambridge: Polity, 2006).

58. The "segregation" of the internet and the self-reinforcing effect of its "echo chambers" were already identified by Cass R. Sunstein, *Republic.com 2.0. Revenge of the Blogs* (Princeton, NJ: Princeton University Press, 2007), 149, 144; Eli Pariser, *The Filter Bubble: What the Internet Is Hiding from You* (New York: Penguin, 2011); and Ethan Zuckerman, *Rewire: Digital Cosmopolitans in the Age of Connection* (New York: Norton, 2013). See also the discussion in chapter 1.

59. Vilém Flusser, "Wohnung Beziehen in der Heimatlosigkeit" [Making a home in homelessness], *Du, Die Zeitschrift der Kultur* 12 (1992): 14.

60. That information is superior to noise in its capacity for differentiation is the core of Gregory Bateson's famous definition of information as the difference (in relation to previous knowledge) that makes a difference (for future actions). Gregory Bateson, *Steps to an Ecology of Mind* (Chicago: University of Chicago Press, 1972), 453.

61. Flusser, "Wohnung Beziehen in der Heimatlosigkeit," 14; Vilém Flusser, *The Freedom of the Migrant: Objections to Nationalism*, trans. Kenneth Kronenberg (Urbana: University of Illinois Press, 2003). Flusser's euphoric grasp of the emblematic function of the migrant for contemporary society differs both from Giorgio Agamben's problematization of the refugee as the *homo sacer* of the present and from Benedict Anderson's diaspora or "long-distance nationalism."

62. Jon Katz, "Birth of a Digital Nation," *Wired*, April 5, 1997, http://archive.wired.com/wired/archive/5.04/netizen_pr.html.

63. Mark Poster, "Digital Networks and Citizenship," *PMLA* 117, no. 1 (January 2002): 100, 102.

64. Milton Mueller, "Internet Nation?," 2014, http://www.internetgovernance.org/2014/09/05/internet-nation.

65. "Bitcoin has even shown that we can have a global, non-state currency through digital technology." Mueller, "Internet Nation." The Swiss firm Swatch and MIT professor Nicholas Negroponte tried in 1998 to

create an internet time that would replace the system of 24 hours of 60 minutes and 60 seconds each with 1,000 beats of one minute and 26.4 seconds each, marked by the @ sign. Internet time was meant to have an existence independent of the normal time zones, so that it would be @500 in Berlin at the same time as in Hong Kong and Rio de Janeiro.

66. Gerald Delanty, *Community* (London: Routledge, 2010), writes about "thin universalistic identity" and "thin" (virtual) communities, whose existence depends on making communication an "essential feature of belonging" (132, 137, 135).

67. Arjun Appadurai, "Archive and Aspiration," in *Information Is Alive*, ed. Joke Brouwer and Arjen Mulder (Rotterdam: V2, 2003), 17. On the WELL as a "thick virtual community," see Delanty, *Community*, 141; and Howard Rheingold, *The Virtual Community: Surfing the Internet* (New York: Minerva, 1995).

68. Peter van Ham, "Europe's Postmodern Identity: A Critical Appraisal," in *Global Society in Transition: An International Politics Reader*, ed. Daniel N. Nelson and Laura J. Neack (New York: Kluwer Law International, 2002), 200. Van Ham's argument is directed at Anthony Smith's view that memory is central for identity formation and that therefore "existing 'deep' cultures" cannot be replaced by "a cosmopolitan 'flat' culture" (192).

69. Ran Zwigenberg: *Hiroshima: The Origins of Global Memory Culture* (Cambridge: Cambridge University Press, 2014). On the concept of the "unbounded universal 'we'" see Levy and Sznaider, *Human Rights and Memory*, 1. The European Union, it is true, derives its motivation and founding precisely from the memory of the catastrophic events of European history. Weinrich, in this sense, makes note of an "ancient enmity" (*Urfeindschaft*) between morality and forgetting, which results in replacing order by contingency. Harald Weinrich, *Gibt es eine Kunst des Vergessens?* [Is there an art of forgetting?] (Basel: Schwabe, 1996), 48.

70. Siegfried Zielinski, *[. . . After the Media]: News from the Slow-Fading Twentieth Century*, trans. Gloria Custance (Minneapolis, MN: Univocal, 2013), 244 (translation modified). Zielinski mentions the concept of "dis-membering" (*Entinnern*) only in passing and refers to Klaus Bartels, "Erinnern, Vergessen, Entinnern. Das Gedächtnis des Internet" [Recalling, forgetting, dis-membering: The memory of the internet], In *Lab—Jahrbuch 2000 für Künste und Apparate* [Lab—Yearbook 2000 for arts and apparatuses], ed. Thomas Hensel, Hans Ulrich Reck, and Siegfried Zielinski (Cologne: Verlag der Buchhandlung Walther

König, 2000), 7–16. Bartels himself borrows the term from Wolfram Malte Fues: "Re-membering, Dis-membering: Fictionality and Hyperfictionality," in *The Poetics of Memory*, ed. Thomas Wägenbauer (Tübingen: Stauffenburg, 1998), 391–98.

71. Bartels, "Erinnern, Vergessen, Entinnern," 12–13. Bartels explains the method of the Cistercians in the context of Janet Coleman, "Das Bleichen des Gedächtnisses. Hl. Bernhards monastische Mnemotechnik" [The bleaching of memory: St. Bernhard's mnemotechnics], in *Gedächtniskunst: Raum-Bild-Schrift. Studien zur Mnemotechnik* [The art of memory: Space—image—script: Studies on mnemotechnics], ed. Anselm Haverkamp and Renate Lachmann (Frankfurt: Suhrkamp, 1993), 207–27. Bartels applies the concept of dis-membering to role-playing in internet chats, where the assumed role of a media icon (Jimmy Stewart or Katherine Hepburn, for example) is determined by the cultural schematics of the role itself. The conclusion seems to be overly hasty, since role-playing in chats is controlled by the individuals in question and remains limited in time. Nor does the concluding equivalence Bartels claims between dis-membering and the loss of any sure sense of reality in David Cronenberg's sciencefiction film *eXistenZ* support his thesis that "the WWW is not only not a memory, it 'dis-members' memory" (10).

72. Jean-Luc Nancy, *Being Singular Plural*, trans. Robert Richardson and Anne O'Byrne (Stanford, CA: Stanford University Press, 2000), xxii.

73. Jean-Luc Nancy, *The Inoperative Community*, trans. Peter Connor et al. (Minneapolis: University of Minnesota Press, 1991), 12. See also Nancy, *Being Singular Plural*: "*Being itself is given to us as meaning. Being does not have meaning. Being itself, the phenomenon of Being, is meaning that is, in turn, its own circulation—and we are this circulation. There is no meaning if meaning is not shared, and not because there would be an ultimate or first signification that all beings have in common, but because meaning is itself the sharing of Being*" (2). The political reference point for Nancy's model of community is the nationalist and ethnic conflicts and acts of violence of the 1990s; the philosophical reference point is Bataille, who, similarly, developed his idea of a community without (post-Christian) *communio* during the 1930s, at a time when communism and fascism appeared to offer the two great seductive and merciless versions of community.

74. Nancy, *Being Singular Plural*, 87. Further: "It is not enough, then, to set idle chatter in opposition to the authenticity of the spoken word,

understood as being replete with meaning. On the contrary, it is necessary to discern the conversation (and sustaining) of being-with as such within chatter: it is in 'conversing,' in the sense of discussion, that being-with 'sustains itself,' in the sense of the perseverance in Being. . . . In this conversation (and sustaining) of being-with, one must discern how language, at each moment, with each signification, from the highest to the lowest—right down to those 'phatic,' insignificant remarks ('hello,' 'hi,' 'good' . . .) which only sustain the conversation itself—exposes the with, exposes itself as the with, inscribes and *ex-scribes* itself in the with until it is exhausted, emptied of signification" (87). In this sense, Dallmayr notes, in regard to Nancy's concept of community, "What is involved in this originary society is neither fusion nor exclusion, but a kind of 'communication' that is vastly different from a mere exchange of information or messages. In opposition to technical information theories (and also theories of communicative interactions), Nancy locates communication on a more primary level; that of the 'sharing and . . . com-pearance (*com-parution*) of finitude.'" Fred Dallmayer, "An 'Inoperative' Global Community? Reflections on Nancy," in *On Jean-Luc Nancy: The Sense of Philosophy*, ed. Darren Sheppard, Simon Sparks, and Colin Thomas (London: Routledge, 1997), 181. Dallmayer's quotation from Nancy is from *The Inoperative Community*, 29.

75. See Frank Vetere, Steve Howard, and Martin R. Gibbs, "Phatic Technologies: Sustaining Sociability Through Ubiquitous Computing," in *Proceedings of the CHI-Conference 2005*, http://www.vs.inf .ethz.ch/events/ubisoc2005/UbiSoc%202005%20submissions/12 -Vetere-Frank.pdf; Victoria Wanga, John V. Tuckera, and Kevin Haines, "Phatic Technologies in Modern Society," *Technology in Society* 33, no. 1 (2012): 84–93. On "disinterested interest" and "pan-sympathy" with a nod to Hume, see Abrol Fairweather and Jodi Halpen, "Do Status Updates Have Any Value?" in *Facebook and Philosophy: What's on Your Mind?*, ed. Dylan E. Wittkower (Chicago: Open Court, 2010), 193, 195, 196. The authors also emphasize that the sympathy created by status updates does not automatically translate into morally significant empathy, for which a deeper encounter with the other would be necessary than is offered by the "ambient awareness" of status updates (198–99). We will come back to this difference.

76. On "singular beings," see Nancy, *The Inoperative Community*, 27; on the "ecstasy of sharing," 25. Further, "Ecstasy . . . implies no effusion,

and even less some form of effervescent illumination." Rather, it should be understood as the "impossibility . . . of absolute immanence" (6). When Nancy later notes, in reference to Bataille's concept of lovers, that "the sovereignty of the lovers is no doubt nothing other than the ecstasy of the instant, it does not *produce a union*, it is NOTHING—but this nothing itself is also, in its 'consummation,' a communion" (37), this passage recalls Pschera's apologia for Facebook as a network that allows its users to be "lovers of the moment" who transcend purposefulness. Alexander Pschera, *800 Millionen. Eine Apologie der sozialen Medien* [800 million: An apologia for social media] (Berlin: Matthes & Seitz, 2011). On Nancy's distancing from a community that "realizes itself as a work," see 8.

77. Nancy, *Inoperative Community*, 27. In a certain sense, Nancy's model is close to Maffesoli's concept of "emotional community," which also has no other ground than that of communality but which relies strictly on loyalty and conformity. Michel Maffesoli, *The Time of the Tribes: The Decline of Individualism in Mass Society*, trans. Don Smith (London: Sage, 1996). For concepts of inessential commonality and solidarity without closure after Nancy and in response to him, see Giorgio Agamben, *The Coming Community* (Minneapolis: University of Minnesota Press, 1993); and Alphonso Lingis, *The Community of Those Who Have Nothing in Common* (Bloomington: Indiana University Press, 1994). Both conceive community as becoming and constant reworking rather than as identity and belonging. For an attempt to carry Nancy's theory over into real contexts, see the chapter "Community of Dissensus" in Bill Readings, *The University in Ruins* (Cambridge, MA: Harvard University Press, 1996), which operates with Nancy's and Maurice Blanchot's concept of a "community without identity" (185).

78. Tiqqun, *Theory of Bloom*, trans. Robert Hurley (London: LBC, 2012), 44. Tiqqun identifies the overcoming of cultural concreteness as "rootless" (46), on the one hand, in the context of Karl Marx's theory of exchange value, according to which the isolated individual, liberated from all traditional social relations of dependency, finds a social context only via the exchange value of labor and commodities. But at the same time, it also interprets this rootlessness, wholly in Nancy's sense of insubstantial community, as the end of the "falsity of membership to a class, to a nation, to a milieu. . . . Only a radical alienation of the Common was able to hypostatize the originary Common

in such a way that solitude, finitude, and exposure, that is, the only actual connection between men, also appears as the only possible connection between them" (54–55). Only the fact that they have been robbed of the content of life, accordingly, "qualifies" subjects as humans per se, who are free to connect with other (equally uprooted) "singularities," to use Nancy's term.

79. Jean-Luc Nancy, "Of Struction," *Parrhesia* 17 (2013), http://www.parr hesiajournal.org/parrhesia17/parrhesia17_nancy.pdf. See also Erich Hörl, "Die künstliche Intelligenz des Sinns. Sinngeschichte und Technologie im Anschluss an Jean-Luc Nancy" [The artificial intelligence of sense: History of sense and technology following Jean-Luc Nancy], *Zeitschrift für Medien- und Kulturforschung* 2 (2010): 135. The concepts "technological shift of meaning" and "cybernetic subjectivity" are drawn from this essay (133) and its introduction (33).

80. Katz, "Birth of a Digital Nation." See also Mark Zuckerberg's manifesto "Is Connectivity a Human Right?" (http://www.facebook.com /isconnectivityahumanright; https://fbcdn- dragon-a.akamaihd.net /hphotos-ak-ash3/851575_2287942339372 24_51579300_n.pdf) and the call that was initiated in September 2015 with ONE #connecttheworld (http://connecttheworld.one.org), as well as Zuckerberg's speech to the UN on September 26, 2015, on the significance of universal access to the internet for information, exchange of ideas, political participation, and job opportunities.

81. Lee Raine and Aaron Smith, "Social Networking Sites and Politics," *PEW Reports*, March 12, 2012, http://www.pewinternet.org/~/media /Files/Reports/2012/PIP_SNS_and_politics.pdf; Lisa Yuk-ming Leung, "Intimacy for 'Deliberative Democracy'? The Role of 'Friendship' in the Participatory Use of Facebook for Activists in Hong Kong," paper presented at the eighth annual conference of the Asian Studies Association of Hong Kong, Hong Kong Institute of Education, China, March 2013. On the internet as a "dialectical space," see Christian Fuchs, *Foundations of Critical Media and Information Studies* (London: Routledge, 2011), 291: "One should therefore better not speak of the contemporary web 2.0 as the 'participatory web 2.0.' but as the web of exploitation and exclusion." Fuchs refers to Herbert Marcuse on the "repressive tolerance" of the "corporate web 2.0" as a device that affords maximal freedom of expression with minimal social effects due to a lack of visibility (276). See also his chapter "Alternative Media as Critical Media" (295–322). On internet

networks as "commodification of freedom," see Manuel Castells, *Communication Power* (Oxford: Oxford University Press, 2009), 421.

82. Hossein Derakhshan, "Das Internet, das wir bewahren müssen" [The internet that we need to preserve], *Die Zeit Online*, July 22, 2015, http://www.zeit.de/digital/internet/2015-07/social-media-blogger-iran-gefaengnis-internet. That the finding of banality is not (yet) accurate everywhere is demonstrated by news reports on political bloggers in China, Bangladesh, and elsewhere who are being arrested by state agencies and murdered by fundamentalist gangs.

83. Hubert L. Dreyfus, *On the Internet* (London: Routledge, 2001), 73, 89, 102–3.

84. Morozov compares and contrasts a person's membership in Facebook groups like Saving-Darfur with nonparticipation in political committees at his own university. Evgeny Morozov, *The Net Delusion: The Dark Side of Internet Freedom* (New York: Public Affairs, 2011), 194. Dean criticizes the phenomenon of mere "registration of political statements on social media like MySpace and Facebook using Slavoj Žižek's concept of "interpassivity," as a wild actionism that despite its interactivity actually prevents activity. Jodi Dean, *Democracy and Other Neoliberal Fantasies: Communicative Capitalism and Left Politics* (Durham, NC: Duke University Press, 2009), 31. A new form of half-hearted engagement is the German app Goodnity (http://goodnity.com), which finances "Adopt a Child" programs worldwide by having users answer marketing questions or look at advertising on their smartphone. The money raised by the app is sent to aid organizations. In this way, say the developers of the app, "doing good" is firmly anchored in the everyday. An unfriendly interpretation might read: social engagement by means of spending time in the service of consumer culture. It is, therefore, hardly surprising to find that two years later the app turned out to be a service for employers to improve their corporate culture. The opposite example is Anonymous, as a protest culture that also takes place on a person's own screen but that has clearly been radicalized and politicized and also brings the protest (as "Operation Chanology," against Scientology in early 2008 demonstrates) into the street.

85. Dreyfus seems to wish for this sort of "liberation" from despair at the noncommittal nature of things, without seeing it as particularly likely (*On the Internet*, 87).

86. See, for instance, Matthias Alexander's response to the survey "Weckruf: Studenten, was geht?" [Wake up, students, what's going on here?] in the *Frankfurter Allgemeine Zeitung*, on the mental condition of students, on July 17, 2014: "If they had an opponent it could keep them from blathering on" (http://www.faz.net/aktuell/beruf-chance /campus/weckruf-an-die-aktuelle-studentengeneration-13039149-p3 .html).

87. "Those who are bragging about their ethics and their humanity today are only waiting to persecute those they condemn by their criteria." Theodor Adorno, "Commitment," in *Notes to Literature*, ed. Rolf Tiedemann, trans. Shierry Weber Nicholsen (New York: Columbia University Press, 1992), 2:92–93.

88. For a critique of Nancy's "philosophism" as an "attempt to think being-with from within philosophy alone" and as "underestimating the constitutive role of conflict and antagonism," see Oliver Marchart, *Post-Foundational Political Thought: Political Difference in Nancy, Lefort, Badiou, and Laclau* (Edinburgh: Edinburgh University Press, 2007), 80, 81. As Marchart emphasizes, the "we" of every community comes from "a homogenizing construction out of the dispersed plurality of being," with the necessary consequence that every concept of community will always need "*some* foundation" (81). In denying this constructing, "antagonistic" moment of the political, Marchart claims that Nancy was committing himself to a "depoliticized notion of the political" that replaces the "fundamentalism of the ground with the fundamentalism of *no ground*" (82).

89. Hanna Meretoja, *The Narrative Turn in Fiction and Theory: The Crisis and Return of Storytelling from Robbe-Grillet to Tournier* (New York: Palgrave Macmillan, 2014), 207, 229. It is obvious that such self-critical metanarratives are fundamentally different from Lyotard's (legitimating) metanarratives (or grand narrative), which give a totalizing (and reductionist) account of historical and cultural phenomena appealing to a universal truth. In the realm of cinematic narration, Meretoja's counterpart would be Peter Greenaway, who declared, "I take no position. I believe that there are no more positions to take, no certainties, no facts. Many people find this very confusing in my films, they say you are hiding behind your irony." Andreas Kilb, "Peter Greenaway oder Der Bauch des Kalligraphen" [Peter Greenaway or the belly of the calligrapher], in *Die Postmoderne im Kino. Ein Reader*

[Postmodernism in the cinema: A reader], ed. Jürgen Felix (Marburg: Schüren, 2002), 235.

90. Meretoja (*The Narrative Turn*, 212) cites the "'weakened' experience of truth" from Gianni Vattimo, *The Transparent Society*, 42. More pertinent is Gianni Vattimo, *Beyond Interpretation: The Meaning of Hermeneutics for Philosophy*, trans. David Webb (Stanford, CA: Stanford University Press, 1997), where he refers to both the "nihilistic vocation of hermeneutics" (chap. 1) and its "anti-metaphysical orientation" (27).

91. Vattimo, *Beyond Interpretation*, 40. Vattimo's idea recalls Nancy in his suggestion that cosmopolitanism is profoundly postmodern and should be thought not (as in more recent approaches) as an extension of loyalty from neighborhood or nation to larger realms but instead as the overcoming of loyalty itself—beginning with a certain "disloyalty" toward oneself.

92. Martin Heidegger, *Being and Time*, trans. John Macquarrie and Edward Robinson (San Francisco: Harper, 1962), 206. Vattimo, *The Transparent Society*, 9. In a way, the model of tolerance based on indifference recalls the cynic Diogenes of Sinope, whose cosmopolitanism (he was the first to use the term) resembled a mocking rejection of emotional connection to one's own polis more than the emphatic relationship to the cosmos that later defined the Stoics' concept of cosmopolitanism.

93. This is the view expressed by Beck, *Cosmopolitan Vision*. It is evident that the concept of cosmopolitanism, in its parallelism to "weak thinking" and in the context of social networks, should not be conceived either as world citizenship beyond nation and region (often viewed as imperialistic and arrogant) or as a hybrid à la "rooted cosmopolitanism." Rather, it should be understood as (self-critical) openness to the Other that is different from the polis, neighborhood, or "home" of the I—including the thought and value system of another Facebook user. It is worth noting, in the relevant debate, that precisely those writers who present this understanding of cosmopolitanism (as a "search for contrast rather than uniformity") tend to refer to Berlin's thought model of the fox. For example, Hannerz writes that "cosmopolitans should ideally be foxes rather than hedgehogs." Ulf Hannerz, "Cosmopolitans and Locals in World Culture," *Theory, Culture & Society* 7 (1990): 239.

94. "Community-based art" resembles social networks to the extent that it is also concerned less with comparing perspectives, positions, and interpretations than with the creation of social spaces that enable

communication among an interactive "public." The curator and art theoretician Nicolas Bourriaud, *Relational Aesthetics* (Dijon: Les Presses du Réel, 2002), 44, cites Rirkrit Tiravanija's cooking performances as paradigmatic for these "alternative forms of sociability" and "moments of constructed conviviality." Leftist art theory criticizes the "shaky analogy between an open work and an inclusive society, as if a desultory form might evoke a democratic community, or a non-hierarchical installation predict an egalitarian world" that it finds at the heart of Bourriaud's aesthetics. Hal Foster, "Chat Rooms," in *Participation: Documents of Contemporary Art*, ed. Claire Bishop (Cambridge, MA: MIT Press, 2006), 193. Commentators take offense, among other things, at the "feel-good position" adopted by Tiravanija: "In such a cozy situation, art . . . collapses into compensatory (and self-congratulatory) entertainment." Claire Bishop, "Antagonism and Relational Aesthetics," *October* 110 (Fall 2004): 79. Referring to political theory's "concept of antagonism," as developed in Ernesto Laclau and Chantal Mouffe, *Hegemony and Socialist Strategy: Towards a Radical Democratic Politics* (London: Verso, 1985). Bishop emphasizes that "a democratic society is one in which relations of conflict are *sustained*, not erased," (66–67). Bishop favors a "relational antagonism" that does not claim social harmony but rather exposes the tensions that are repressed by the appearance of harmony (79). Naturally, "feel-good" get-togethers (a cooking project by Tiravanija, or Cyprien Gaillard's 2011 "Recovery of Discovery" in the Berlin art project: a pyramid of beer bottles that the public was invited to deconstruct by drinking them) can also incorporate social-critical features, for example when the conversation turns to those who have been excluded from the party. Facebook, too, for all its rhetoric of togetherness, makes relational antagonism possible by creating space where opposing positions can clash. The core of the critique is the differing orientation: the intent to avoid conflict versus the attempt to make it productive. The core issue, for this tendency, is the difference in hermeneutical approach. Relational aesthetics does not aim to create a meaningful work and work-based interpretation; instead, it seeks to create a social situation and shared experience of that situation. The opposing position (which Bishop illustrates by drawing on interaction artists Santiago Sierra and Thomas Hirschhorn) aims at interpreting the created situation and resists understanding it at the level of a party: "The most important activity that an art work can provoke is the activity of thinking," and "having reflections and critical thoughts is to get active." Hirschhorn, cited in Bishop, *Participation*, 76–77. Surprisingly, Bishop locates the core meaning of Nancy's community concept in its "counter-model to relational aesthetics" (68), and she includes passages from *The Inoperative Community* in her *Participation* reader. However, she does not explain there, or in her later book *Artificial Hells: Participatory Art and the Politics of Spectatorship* (London: Verso, 2012), how Nancy's model of a society with no substance can be mobilized against an artistic practice of superficial sociality. Baker is more clear in this regard, when, in reference to Nancy, he argues for an aesthetics that, in contrast to the "cynical aesthetics of immediate community," thematizes the "break or fissure in social orders and social groups." George Baker, "Beziehungen und Gegenbeziehungen. Ein offener Brief an Nicolas Bourriaud" [Relationships and counter-relationships: An open letter to Nicolas Bourriaud], in *Contextualize: Zusammenhänge herstellen* [Contextualize: Create contexts], ed. Yilmaz Dziewior (Cologne: DuMont Buchverlag, 2003), 128. In my study *Digital Art and Meaning: Reading Kinetic Poetry, Text Machines, Mapping Art, and Interactive Installations* (Minneapolis: University of Minnesota Press, 2011), I explore the relationship of participatory art to reflection and interpretation, with reference to Bourriaud (122–57). A detailed analysis of the psychological and political parallels between participation art and participation culture would be a task for future research on Facebook.

AFTERWORD

1. Siegfried Kracauer, "Photography," trans. Thomas Y. Levin, *Critical Inquiry* 19, no. 3 (Spring 1993): 432, 435, 436, 434 (translation modified).
2. Benjamin, however, seems to have been convinced of the good outcome of history's game of chance when he praised the "men who have adopted the cause of the new and have founded it on insight and renunciation": "In its buildings, pictures, and stories, mankind is preparing to outlive culture, if need be. And the main thing is that it does so with a laugh." Walter Benjamin, "Experience and Poverty," in *Selected Writings*: vol. 2, part 2: *1931–1934*, ed. Michael W. Jennings, Howard Eiland, and Gary Smith (Cambridge, MA: Harvard University Press, 1999), 732, 735. The following quotation on the "foundation of nature devoid of meaning" is from Kracauer, "Photography," 434.

3. The concept of "distant reading," promoted by Franco Moretti, *Distant Reading* (London: Verso, 2013), means the algorithmic, keyword-oriented analysis of large numbers of texts, as distinguished from interpretation by means of close reading. Anderson proclaimed the end of theory in 2008, explaining that "with enough data, the numbers speak for themselves." Chris Anderson, "The End of Theory: The Data Deluge Makes the Scientific Method Obsolete," *Wired*, June 23, 2008, http://www.wired.com/ science/discoveries/magazine/16-07 /pb_theory. Ramsay argues that algorithmic analyses do not have to end in the positivism of "objective" statements but can lead to new questions for research that may remain quite open to a multiperspectival interpretation. Stephen Ramsay, "Toward an Algorithmic Criticism," *Literary and Linguistic Computing* 18, no. 2 (2003): 167–74. On the concept of an "ecology of collaborating," Hayles remarks, "The humanities cannot continue to take the quest for meaning as an unquestioned premise for their ways of doing business." N. Katherine Hayles, "Cognition Everywhere: The Rise of the Cognitive Nonconscious and the Costs of Consciousness," *New Literary History* 45, no. 2 (2014): 217, 199. I discuss the relationship of the humanities to digital media extensively in my book *Medien und Bildung* [Media and literacy] (Berlin: Matthes & Seitz, 2018). For my view on digital humanities, see Roberto Simanowski and Luciana Gattass, "Debates in the Digital Humanities Formerly Known as Humanities Computing," *electronic book review*, March 5, 2017, http://electronicbookreview.com /thread/electropoetics/debated.

4. On "transformation of the human," see Michael Hagner and Erich Hörl, eds., *Die Transformation des Humanen. Beiträge zur Kulturgeschichte der Kybernetik* [The transformation of the human: Contributions to a cultural history of cybernetics] (Berlin: Suhrkamp, 2008); on "nonconscious cognition" and "distributed cognition environments," see Hayles, "Cognition Everywhere," and N. Katherine Hayles, *How We Became Posthuman: Virtual Bodies in Cybernetics, Literature, and Informatics* (Chicago: University of Chicago Press, 1999).

5. All quotations are from Erich Hörl's lecture at MECS (Institute for Advanced Study on Media Cultures of computer Simulation), Lüneburg, July 2014: "Milieus der Modulation. Zur Aktualität von Gilbert Simondons spekulativer Ökologie" [Milieus of moderation: On the contemporary relevance of Gilbert Simondon's speculative ecology], http://www.youtube.com/watch?v=GehfVn-MYJM, approx. 20 min.

On the "fourth insult," see Hagner and Hörl, eds., *Die Transformation des Humanen*, 10. The equivalency posed there between the cybernetic "insult" and Foucault's "death of man" ignores the psychological difference between the insight that man is not the sovereign source of his thoughts (or the center of the universe) and the sovereign transfer of thinking to other actants. The latter is an insult, a "technological insult" (beyond the trivial insult of the power differential among humans, for example, between programmers, controllers, and objects of a surveillance technology) only as an accident, if a person underestimates the independent dynamics of a system that has been set in motion and loses control over the "spirits" he has summoned, as Goethe's sorcerer's apprentice did. Johannes Rohbeck, *Technologische Urteilskraft—Zu einer Ethik technischen Handelns* [Technological judgment: Toward an ethics of technical action] (Frankfurt: Suhrkamp, 1993), 10. The following quotations on the "correction" of "cybernetics" are drawn from Hörl's lecture.

6. On the quotation from Nancy, see chap. 3, note 74. The opposing perspectives that are proposed in German-language media studies on the subject of the current go-for-broke game can be distilled down to the dispute between Erich Hörl and Dieter Mersch, who negotiated their understandings of the technological present (on cybernetics, ecologies, and the rule of mathematics) in a conversation held on November 6, 2014, at the Berlin Akademie der Künste: http://www.youtube .com/watch?v=PoZ3GjVXcFo. For Mersch, who, contradicting Hörl, traces the technological condition back to its mathematical foundation, the cybernetic disempowerment of humanity by no means contains a sly dialectics but only an "'antihumanistic' impulse" accompanied by a renunciation of sovereignty, of humankind's hegemonic power over the objects of this world, and thus also over itself. As Mersch emphasizes, "We have, then, to do with alternative theaters of the social, in which technology occupies a place that is equal to, or enjoys equal rights alongside other cultural formations, without being dominated by goals or rules of engagement, which alone would seem to be adequate for 'orders' given by humankind." Dieter Mersch, *Ordo ab chao—Order from Noise* (Zurich: Diaphanes, 2013), 17, 18. Precisely this loss of human dominance over humanity and the alternative regulation of the social are, it should be noted, the promise that would be to be realized from the opposite perspective on the new go-for-broke game's technological constellation. A media studies expert with a

literary background, it should be noted, is already made unhappy by taking the game of chance as a point of departure and naturally hopes that the solution of the problem (Nancy's problem and all the problems of humankind) will be found in the model of the narrative, rather than in mathematics.

7. On right and left cybernetics, see Mersch, *Ordo ab chao*, 77–85.

EPILOGUE TO THE ENGLISH EDITION

1. Mat Honan, "Why Facebook and Mark Zuckerberg Went All In on Live Video," *BuzzFeed*, April 6, 2016, http://www.buzzfeed.com /mathonan/why-facebook-and-mark-zuckerberg-went-all-in-on-live -video?utm_term=.vh7baAPxeN#.cwkDwVNX37.

2. http://www.facebook.com/notes/mark-zuckerberg/building-global -community/10154544292806634.

BIBLIOGRAPHY

Adorno, Theodor W. *Minima Moralia: Reflections from Damaged Life.*
Trans. Edmund Jephcott. New York: Verso, 2006.

Adorno, Theodor W. "Commitment." In *Notes to Literature*, ed. Rolf Tiedemann, trans. Shierry Weber Nicholsen, 2:76–94. New York: Columbia University Press, 1992.

Adorno, Theodor W., and Max Horkheimer. *Dialectic of Enlightenment.*
Ed. Gunzelin Schmid Noerr. Trans. Edmund Jephcott. Stanford, CA:
Stanford University Press, 2002.

Agamben, Giorgio. *The Coming Community.* Minneapolis: University of
Minnesota Press, 1993.

——. *Infancy and History: On the Destruction of Experience.* Trans. Liz
Heron. New York: Verso, 1993.

——. "What Is an Apparatus?" In *"What Is an Apparatus?" and Other Essays*,
trans. David Kishik and Stefan Petadella, 1–24. Stanford, CA: Stanford
University Press, 2009.

——. "What Is the Contemporary?" In *"What Is an Apparatus?" and Other
Essays*, trans. David Kishik and Stefan Petadella, 39–54. Stanford, CA:
Stanford University Press, 2009.

Albrechtslund, Anders, and Lynsey Dubbeld. "The Plays and Arts of Surveillance: Studying Surveillance as Entertainment." *Surveillance & Society* 3, no. 2/3 (2005): 216–21.

Anderson, Benedict. *Imagined Communities: Reflections on the Origin and
Spread of Nationalism.* New York: Verso, 1996.

——. "The New World Disorder." *New Left Review* 193 (1993): 3–13.

Anderson, Chris. "The End of Theory: The Data Deluge Makes the Scientific
Method Obsolete." *Wired*, June 23, 2008, http://www.wired.com/science
/discoveries/magazine/16-07/pb_theory.

Appadurai, Arjun. "Archive and Aspiration." In *Information Is Alive*, ed. Joke Brouwer and Arjen Mulder, 14–25. Rotterdam: V2, 2003.

——. *The Future as Cultural Fact: Essays on the Global Condition.* London: Verso, 2013.

Aronowitz, Stanley. "Looking Out: The Impact of Computers on the Lives of Professionals." In *Literacy Online: The Promise (and Peril) of Reading and Writing with Computers*, ed. Myron C. Tuman, 119–37. Pittsburgh, PA: University of Pittsburgh Press, 1992.

Assmann, Aleida. "Hier bin ich, wo bist du? Einsamkeit im Kommunikationszeitalter" [Here I am, where are you? Loneliness in the age of communication]. *Mittelweg* 36, no. 1 (2011): 4–23.

——. *Memory and Political Change.* Trans. Linda Shortt. New York: Palgrave Macmillan, 2012.

Baker, George. "Beziehungen und Gegenbeziehungen. Ein offener Brief an Nicolas Bourriaud" [Relationships and counter-relationships: An open letter to Nicolas Bourriaud]. In *Contextualize: Zusammenhänge herstellen* [Contextualize: Create contexts], ed. Yilmaz Dziewior, 126–133. Cologne: DuMont Buchverlag, 2003.

Balász, Béla. *Early Film Theory: Visible Man and the Spirit of Film.* Trans. Rodney Livingstone. New York: Berghahn, 2011.

Ball, Hugo. *Flight Out of Time: A Dada Diary.* Ed. John Elderfield. Trans. Ann Raimes. Berkeley: University of California Press, 1996.

Bartels, Klaus. "Erinnern, Vergessen, Entinnern. Das Gedächtnis des Internet" [Recalling, forgetting, dis-membering: The memory of the internet]. In *Lab—Jahrbuch 2000 für Künste und Apparate* [Lab—Yearbook 2000 for arts and apparatuses], ed. Thomas Hensel, Hans Ulrich Reck, and Siegfried Zielinski, 7–16. Cologne: Verlag der Buchhandlung Walther König, 2000.

Barthes, Roland. *Camera Lucida: Reflections on Photography.* Trans. Richard Howard. New York: Hill and Wang, 1981.

——. *Writing Degree Zero.* Trans. Annette Lavers and Colin Smith. New York: Hill and Wang, 1968.

Bateson, Gregory. *Steps to an Ecology of Mind.* Chicago: University of Chicago Press, 1972.

Baudrillard, Jean. *Photographies 1985–1998* [exhibition catalog]. Ed. Peter Weibel. Trans. Susanne Baumann et al. Ostfildern-Ruit: Hatje Cantz, 1999.

Bauman, Zygmunt. *Liquid Life.* Cambridge, MA: Polity, 2005.

——. *Liquid Love: On the Frailty of Human Bonds.* Cambridge, MA: Polity, 2003.

——. "From Pilgrim to Tourist—or a Short History of Identity." In *Questions of Cultural Identity*, ed. Stuart Hall and Paul du Gay, 18–36. London: Sage, 1996.

——. "Privacy, Secrecy, Intimacy, Human Bonds, Utopia—and Other Collateral Casualties of Liquid Modernity." In *Modern Privacy: Shifting Boundaries, New Forms*, ed. Harry Blatterer, Pauline Johnson, and Maria R. Markus, 7–22. New York: Palgrave Macmillan, 2010.

Beck, Ulrich. *Cosmopolitan Vision.* Trans. Cioran Cronin. Cambridge, MA: Polity, 2006.

Benda, Julien. *The Treason of the Intellectuals.* Trans. Richard Aldington. New Brunswick, NJ: Transaction, 2007.

Benhabib, Seyla. *Claims of Culture: Equality and Diversity in the Global Era.* Princeton, NJ: Princeton University Press, 2002.

Benjamin, Walter. *The Arcades Project.* Trans. Howard Eiland and Kevin McLaughlin. Cambridge, MA: Harvard University Press, 2002.

——. "Central Park." In *Selected Writings*, vol. 4: *1938–1940*, ed. Howard Eiland and Michael W. Jennings, trans. Edmund Jephcott and Howard Eiland, 161–99. Cambridge, MA: Harvard University Press, 2006.

——. "Experience." In *Selected Writings*, vol. 1: *1913–1926*, ed. Marcus Bullock and Michael W. Jennings, 3–5. Cambridge, MA: Harvard University Press, 2004.

——. "Experience and Poverty." In *Selected Writings*, vol. 2, part 2: *1931–1934*, ed. Michael W. Jennings, Howard Eiland, and Gary Smith, 731–36. Cambridge, MA: Harvard University Press, 1999.

——. "Fragmente zur Sprachphilosophie und Erkenntniskritik" [Fragments on language philosophy and epistemology]. In *Kairos. Schriften zur Philosophie*, ed. Ralf Konersmann, 68–79. Frankfurt: Suhrkamp, 2007.

——. "On Some Motifs in Baudelaire." In *Selected Writings*, vol. 4: *1938–1940*, ed. Howard Eiland and Michael W. Jennings, trans. Harry Zohn, 313–55. Cambridge, MA: Harvard University Press, 2006.

——. "On the Concept of History." In *Selected Writings*, vol. 4: *1938–1940*, ed. Howard Eiland and Michael W. Jennings, trans. Harry Zohn, 389–400. Cambridge, MA: Harvard University Press, 2006.

——. "The Storyteller: Observations on the Works of Nikolai Leskov." In *Selected Writings*, vol. 3: *1935–1938*, ed. Howard Eiland and Michael W. Jennings, 143–66. Cambridge, MA: Harvard University Press, 2002.

——. "The Work of Art in the Era of Its Technological Reproducibility." In *Selected Writings*, vol. 3: *1935–1938*, ed. Howard Eiland and Michael W. Jennings, trans. Edmund Jephcott and Harry Zohn, 101–33. Cambridge, MA: Harvard University Press, 2002.

Bergk, Johann Adam. *Die Kunst, Bücher zu lesen* [The art of reading books]. Jena: Hempel, 1799.

Berlin, Isaiah. *The Hedgehog and the Fox: An Essay on Tolstoy's View of History*. Trans. Henry Hardy. Princeton, NJ: Princeton University Press, 2013.

Berry, David M. *Critical Theory and the Digital*. New York: Bloomsbury, 2014.

Bhaba, Homi K. "Unsatisfied: Notes on Vernacular Cosmopolitanism." In *Text and Nation*, ed. Laura Garcia-Morena and Peter C. Pfeiffer, 191–207. London: Camden House, 1996.

Bishop, Claire. "Antagonism and Relational Aesthetics." *October* 110 (Fall 2004): 51–79.

——. *Artificial Hells: Participatory Art and the Politics of Spectatorship*. London: Verso, 2012.

Bloch, Ernst. *The Principle of Hope*. Trans. Neville Plaice, Stephen Plaice, and Paul Knight. Cambridge, MA: MIT Press, 1986.

Bohrer, Karl Heinz. *Der Abschied: Theorie der Trauer: Baudelaire, Goethe, Nietzsche, Benjamin* [The parting: Theory of mourning: Baudelaire, Goethe, Nietzsche, Benjamin]. Frankfurt: Suhrkamp, 1996.

Bölsche, Wilhelm. *Die naturwissenschaftlichen Grundlagen der Poesie. Prolegomena einer realistischen Ästhetik* [The natural-scientific foundations of poetry: Prolegomena to a realistic aesthetics]. Leipzig: Karl Reissner, 1887.

Bolter, Jay David. "Literature in the Electronic Writing Space." In *Literacy Online: The Promise (and Peril) of Reading and Writing with Computers*, ed. Myron C. Tuman, 19–42. Pittsburgh, PA: University of Pittsburgh Press, 1992.

Bolz, Norbert. *Das konsumistische Manifest* [The consumerist manifesto]. Munich: Wilhelm Fink, 2002.

Börne, Ludwig. "Denkrede auf Jean Paul" [Commemorative address for Jean Paul]. Ed. Karl Rauch. Bern: Franke, 1964.

Bourriaud, Nicolas. *Relational Aesthetics*. Dijon: Les Presses du Réel, 2002.

Boyd, Danah. *It's Complicated: The Social Lives of Networked Teens*. New Haven, CT: Yale University Press, 2014.

Brecht, Bertolt. *Bertolt Brecht on Film and Radio*. Ed. and trans. Marc Silberman. London: Bloomsbury, 2000.

Brown, Wendy. "'The Subject of Privacy': A Comment on Moira Gatens." In *Privacies: Philosophical Evaluations*, ed. Beate Rössler, 133–41. Stanford, CA: Stanford University Press, 2004.

Bruner, Jerome S. "Past and Present as Narrative Constructions." In *Narration, Identity, and Historical Consciousness*, ed. Jürgen Straub, 23–43. New York: Berghahn, 2005.

Buddemeier, Heinz. "Was wird im CyberSpace aus den sozialen Beziehungen?" [What becomes of social relations in cyberspace?]. In *CyberSpace. Virtual Reality, Fortschritt und Gefahr einer innovativen Technik* [Cyberspace: Virtual reality, progress, and danger of an innovative technology], ed. Horst F. Wedde, 31–52. Stuttgart: Urachhaus, 1996.

Burkart, Günter. "When Privacy Goes Public: New Media and the Transformation of the Culture of Confession." In *Modern Privacy, Shifting Boundaries, New Forms*, ed. Harry Blatterer, Pauline Johnson, and Maria R. Markus, 23–38. New York: Palgrave McMillan, 2010.

Butler, Judith. *Giving an Account of Oneself*. New York: Fordham University Press, 2005.

Caeiro, Eduardo, and Fernando Pessoa. *The Keeper of Sheep*. Trans. Edwin Honig and Susan M. Brown. Bronx, NY: Sheep Meadow, 1997.

Carr, Nicholas. *The Shallows: What the Internet Is Doing to Our Brains*. New York: Norton, 2011.

Castells, Manuel. *Communication Power*. Oxford: Oxford University Press, 2009.

Charon, Rita. *Narrative Medicine: Honoring the Stories of Illness*. Oxford: Oxford University Press, 2006.

Chou, Hui-Tzu, and Nicholas Edge. "'They Are Happier and Having Better Lives Than I Am': The Impact of Using Facebook on Perceptions of Others' Lives." *Cyberpsychology, Behavior, and Social Networking* 15, no. 2 (2012): 117–21.

Chouliaraki, Lilie. *The Ironic Spectator: Solidarity in the Age of Posthumanitarianism*. Malden, MA: Polity, 2013.

Cohen, Joshua, ed. *For Love of Country*. Boston: Beacon, 1996.

Coleman, Janet. "Das Bleichen des Gedächtnisses. Hl. Bernhards monastische Mnemotechnik" [The bleaching of memory: St. Bernhard's mnemotechnics]. In *Gedächtniskunst: Raum–Bild–Schrift. Studien zur Mnemotechnik* [The art of memory: Space—image—script: Studies on mnemotechnics], ed. Anselm Haverkamp and Renate Lachmann, 207–27. Frankfurt: Suhrkamp, 1993.

Coupland, Douglas. *Generation X: Tales for an Accelerated Culture.* New York: St. Martin's, 1991.

Coutinho, Lester, Suman Bisht, and Gauri Raje. "Numerical Narratives and Documentary Practices: Vaccines, Targets and Reports of Immunisation Programme." *Economic and Political Weekly* 35, no. 8/9 (February 19–26, 2000): 656–66.

Czarniawska, Barbara. *Narratives in Social Science Research.* London: Sage, 2004.

Dallmayer, Fred. "An 'Inoperative' Global Community? Reflections on Nancy." In *On Jean-Luc Nancy: The Sense of Philosophy*, ed. Darren Sheppard, Simon Sparks, and Colin Thomas, 174–96. London: Routledge, 1997.

Davis, William. "Mark Zuckerberg and the End of Language." *Atlantic*, September 11, 2015. http://www.theatlantic.com/technology/archive /2015/09/silicon-valley-telepathy-wearables/404641.

de Man, Paul. "Autobiography as De-Facement." *MLN* 94, no. 5 (December 1979): 919–30.

Dean, Jodi. *Democracy and Other Neoliberal Fantasies: Communicative Capitalism and Left Politics.* Durham, NC: Duke University Press, 2009.

Delanty, Gerald. *Community.* London: Routledge, 2010.

Deleuze, Gilles. "Control and Becoming." In *Negotiations 1972–1990*, trans. Martin Joughin, 169–76. New York: Columbia University Press, 1997.

——. "Mediators." In *Negotiations 1972–1990*, trans. Martin Joughin, 121–34. New York: Columbia University Press, 1997.

Derakhshan, Hossein. "Das Internet, das wir bewahren müssen" [The internet that we need to preserve]. *Die Zeit Online*, July 22, 2015. http:// www.zeit.de/digital/internet/2015-07/social-media-blogger-iran -gefaengnis-internet.

Deresiewicz, William. "The End of Solitude." *Chronicle of Higher Education*, January 30, 2009. http://chronicle.com/article/The-End-of-Solitude/3708.

Dredge, Stuart. "Facebook Boss Mark Zuckerberg Thinks Telepathy Tech Is on Its Way." *Guardian*, July 1, 2015. http://www.theguardian.com/tech nology/2015/jul/01/facebook-mark-zuckerberg-telepathy-tech.

Dreyfus, Hubert L. *On the Internet.* London: Routledge, 2001.

Dunbar, Robin. *How Many Friends Does One Person Need?: Dunbar's Number and Other Evolutionary Quirks.* Cambridge, MA: Harvard University Press, 2010.

Eakin, Paul John. *Living Autobiographically: How We Create Identity in Narrative*. Ithaca, NY: Cornell University Press, 2008.

Eggers, Dave. *The Circle*. New York: Random House, 2013.

Eisele, Ulf. "Empiristischer Realismus. Die epistemologische Problematik einer literarischen Konzeption" [Empiristical realism: The epistemological problematic of a literary conception]. In *Naturalismus, Fin de siècle, Expressionismus. 1890–1918* [Naturalism, fin de siècle, Expressionism: 1890–1918], ed. York-Gothart Mix, 74–97. Munich: Hanser, 1996.

Eriksen, Thomas Hylland. *Tyranny of the Moment: Fast and Slow Time in the Information Age*. London: Pluto, 2001.

Ernst, Wolfgang. "Das Archiv als Gedächtnisort" [The archive as a site of memory]. In *Archivologie. Theorien des Archivs in Theorie, Medien und Künsten* [Archivology: Theories of the archive in theory, media, and the arts], ed. Knut Ebeling and Stephan Günzel, 177–200. Berlin: Kulturverlag Kadmos, 2009.

——. *Das Gesetz des Gedächtnisses. Medien und Archive am Ende (des 20. Jahrhunderts)* [The law of memory: Media and archive at the end (of the twentieth century)]. Berlin: Kulturverlag Kadmos, 2007.

Esposito, Elena. "Die Formen des Web-Gedächtnisses. Medien und soziales Gedächtnis" [The forms of Web-memory: Media and social memory]. In *Formen und Funktionen sozialen Erinnerns. Sozial- und kulturwissenschaftliche Analysen* [Forms and functions of social remembering: Social and cultural studies analyses], ed. René Lehmann, Florian Öchsner, and Gerd Sebald, 91–103. Wiesbaden: Springer VS, 2013.

——. *Soziales Vergessen. Formen und Medien des Gedächtnisses der Gesellschaft* [Social forgetting: Forms and media of the memory of society]. Frankfurt: Suhrkamp, 2002.

Fairweather, Abrol, and Jodi Halpen. "Do Status Updates Have Any Value?" In *Facebook and Philosophy: What's on Your Mind?*, ed. Dylan E. Wittkower, 191–99. Chicago: Open Court, 2010.

Feenberg, Andrew, and Darin Barney, eds. *Community in the Digital Age: Philosophy and Practice*. Lanham, MD: Rowman & Littlefield, 2004.

Felton, Nicholas. "Numerical Narratives." Lecture at the Department of Design, Media, Arts at UCLA, November 15, 2011. http://video.dma.ucla.edu/video/nicholas-felton-numerical-narratives/387.

Fichte, Johann Gottlob. *Addresses to the German Nation*. Trans. R. F. Jones and G. H. Turnbull. Chicago: Open Court, 1922.

Fischer-Lichte, Erika. *The Transformative Power of Performance: A New Aesthetic*. Trans. Saskya Iris Jain. New York: Routledge, 2004.

Fletcher, Dan. "How Facebook Is Redefining Privacy." *Time*, May 20, 2010. http://content.time.com/time/magazine/article/0,9171,1990798,00.html.

Flusser, Vilém. *The Freedom of the Migrant: Objections to Nationalism*. Trans. Kenneth Kronenberg. Urbana: University of Illinois Press, 2003.

——. "Wohnung Beziehen in der Heimatlosigkeit" [Making a home in homelessness]. *Du, Die Zeitschrift der Kultur* 12 (1992): 12–14.

Forman, Janis. *Storytelling in Business: The Authentic and Fluent Organization*. Stanford, CA: Stanford University, 2013.

Foster, Hal. "Chat Rooms." In *Participation: Documents of Contemporary Art*, ed. Claire Bishop, 190–95. Cambridge, MA: MIT Press, 2006.

Foucault, Michel. *The Care of the Self*. Vol. 3 of *The History of Sexuality*. Trans. Robert Hurley. New York: Random House, 1986.

Fuchs, Christian. *Foundations of Critical Media and Information Studies*. London: Routledge, 2011.

——. *Social Media: A Critical Introduction*. London: Sage, 2014.

Fuchs, Christian, Kees Boersma, Anders Albrechtslund, and Marisol Sandoval, eds. *Internet and Surveillance: The Challenges of Web 2.0 and Social Media*. New York: Routledge, 2012.

Fues, Wolfram Malte. "Re-membering, Dis-membering: Fictionality and Hyperfictionality." In *The Poetics of Memory*, ed. Thomas Wägenbauer, 391–98. Tübingen: Stauffenburg, 1998.

Fukuyama, Francis. *The End of History and the Last Man*. New York: Avon, 1993.

Galloway, Alexander R. "Black Box, Black Bloc." In *Communication and Its Discontents: Contestation, Critique, and Contemporary Struggles*, ed. Benjamin Noys, 237–52. London: Minor Compositions, 2011.

Geissler, Karlheinz. *Lob der Pause. Von der Vielfalt der Zeiten und der Poesie des Augenblicks* [Praise of pauses. On the Diversity of times and the poetry of the moment]. Munich: Oekom, 2012.

Gellner, Ernest. *Nations and Nationalism*. Ithaca, NY: Cornell University Press, 1983.

Gibbs, Samuel. "Facebook Tracks All Visitors, Breaching EU Law." *Guardian*, March 31, 2015. http://www.theguardian.com/technology/2015/mar/31/facebook-tracks-all-visitors-breaching-eu-law-report.

Gillespie, Tarleton. "The Relevance of Algorithms." In *Media Technologies: Essays on Communication, Materiality, and Society*, ed. T. Gillespie, P. J. Boczkowski, and K. A. Foot, 167–95. Cambridge, MA: MIT Press, 2014.

Goethe, Johann Wolfgang von. *Faust*. 2 parts. Trans. Martin Greenberg. New Haven, CT: Yale University Press, 1992, 1998.

Goffman, Erving. *The Presentation of Self in Everyday Life.* New York: Doubleday, 1959.

Gumbrecht, Hans Ulrich. *In 1926: Living at the Edge of Time.* Cambridge, MA: Harvard University Press, 1997.

——. *Our Broad Present: Time and Contemporary Culture.* Trans. Henry Erik Butler. New York: Columbia University Press, 2014.

——. *Production of Presence: What Meaning Cannot Convey.* Stanford, CA: Stanford University Press, 2004.

Hagner, Michael, and Erich Hörl, eds. *Die Transformation des Humanen. Beiträge zur Kulturgeschichte der Kybernetik* [The transformation of the human: Contributions to a cultural history of cybernetics]. Berlin: Suhrkamp, 2008.

Han, Byung-Chul. *Psychopolitics: Neoliberalism and New Technologies of Power.* Trans. Erik Butler. New York: Verso Futures, 2017.

——. *The Scent of Time: A Philosophical Essay on the Art of Lingering.* Trans. Daniel Steuer. New York: Polity, 2017.

Hannerz, Ulf. "Cosmopolitans and Locals in World Culture." *Theory, Culture & Society* 7 (1990): 237–51.

Hayles, N. Katherine. "Cognition Everywhere: The Rise of the Cognitive Nonconscious and the Costs of Consciousness." *New Literary History* 45, no. 2 (2014): 199–220.

——. *How We Became Posthuman: Virtual Bodies in Cybernetics, Literature, and Informatics.* Chicago: University of Chicago Press, 1999.

Hegel, Georg Wilhelm Friedrich. *Aesthetics: Lectures on Fine Art.* 2 vols. Trans. T. M. Knox. Oxford: Clarendon, 1975.

Heidegger, Martin. *Being and Time.* Trans. John Macquarrie and Edward Robinson. San Francisco: Harper, 1962.

Heller, Christian. *Post-Privacy: Prima leben ohne Privatsphäre* [Post-privacy: Living just fine without a private sphere]. Munich: C. H. Beck, 2011.

Herder, Johann Gottfried. "Briefe zu Beförderung der Humanität" [Letters for the advancement of humanity]. In *Herders Sämmtliche Werke,* ed. Bernhard Suphan, vol. 18. Berlin: Weidmann, 1883.

——. *Herders Sämtliche Werke,* vol. 14. Ed. Bernhard Suphan. Berlin: Weidmannsche Buchhandlung, 1877–1913.

Hillebrand, Bruno. *Ästhetik des Augenblicks. Der Dichter als Überwinder der Zeit—von Goethe bis heute* [Aesthetic of the moment: The poet as victor over time—from Goethe to the present]. Göttingen: Vandenhoeck & Ruprecht, 1999.

Hörisch, Jochen. *Wut des Verstehens: Zur Kritik der Hermeneutik* [The rage of understanding: Toward a critique of hermeneutics]. Frankfurt: Suhrkamp, 1988.

Hörl, Erich. "Die künstliche Intelligenz des Sinns. Sinngeschichte und Technologie im Anschluss an Jean-Luc Nancy" [The artificial intelligence of sense: History of sense and technology following Jean-Luc Nancy]. *Zeitschrift für Medien- und Kulturforschung* 2 (2010): 129–47.

——, ed. *Die technologische Bedingung. Beiträge zur Beschreibung der technischen Welt* [The technological condition: Contributions to the description of the technical world]. Berlin: Suhrkamp, 2011.

Humboldt, Wilhelm von. "On the Historian's Task." *History and Theory* 6, no. 1 (1967): 57–71.

Huyssen, Andreas. *Present Past: Urban Palimpsests and the Politics of Memory.* Stanford, CA: University of California Press, 2003.

Ippolita, Geert Lovink, and Ned Rossiter. "The Digital Given: 10 Web 2.0 Theses." *Fibreculture Journal* 14 (2009), http://fourteen.fibreculturejournal .org/fcj-096-the-digital-given-10-web-2-0-theses.

Iyer, Pico. *The Art of Stillness—Adventures in Going Nowhere.* New York: Simon & Schuster, 2014.

Jochum, Uwe. "Die virtuelle Bibliothek" [The virtual library]. In *7 Hügel*, vol. 6: *Wissen* [Knowledge] [catalogue of the exhibition], 35–40. Berlin, 2000.

Johnson, Bobbie. "Privacy No Longer a Social Norm, Says Facebook Founder." *Guardian*, January 11, 2010. http://www.theguardian.com /technology/2010/jan/11/facebook-privacy.

Johnson, Steven. *Everything Bad Is Good for You: How Today's Popular Culture Is Actually Making Us Smarter.* New York: Riverhead, 2005.

Jonas, Hans. *The Imperative of Responsibility: In Search of an Ethics for the Technological Age.* Trans. Hans Jonas and David Herr. Chicago: University of Chicago Press, 1984.

Joyce, Michael. "Nonce Upon Some Times: Rereading Hypertext Fiction." *Modern Fiction Studies* 43, no. 3 (1997): 579–97.

Jurgenson, Nathan. "The Facebook Eye." *Atlantic*, January 12, 2012. http://www .theatlantic.com/technology/archive/2012/ 01/the-facebook-eye/251377.

Kaiser. Eduard, *Trost der Langeweile: Die Entdeckung menschlicher Lebensformen in digitalen Welten* [The consolation of boredom: The discovery of human life forms in digital worlds]. Glarus, Switzerland: Rüegger, 2014.

Kant, Immanuel. "An Answer to the Question: 'What Is Enlightenment?'" Trans. H. B. Nisbet. London: Penguin, 2013.

——. *Groundwork of the Metaphysics of Morals*. Trans. Mary Gregor and Jens Timmermann. Cambridge: Cambridge University Press, 2012.

——. "Idea for a Universal History with a Cosmopolitan Purpose." In *On History*, trans. Lewis White Beck, 11–26. Indianapolis, IN: Bobbs-Merrill, 1963.

Katz, Jon. "Birth of a Digital Nation." *Wired*, April 5, 1997. http://archive .wired.com/wired/archive/5.04/netizen_pr.html.

Kilb, Andreas. "Peter Greenaway oder Der Bauch des Kalligraphen" [Peter Greenaway or the belly of the calligrapher]. In *Die Postmoderne im Kino. Ein Reader* [Postmodernism in the cinema: A reader], ed. Jürgen Felix, 230–38. Marburg: Schüren, 2002.

Kisch, Egon Erwin. *Hetzjagd durch die Zeit* [Feverish hunt through time]. Berlin: Universum Bücherei, 1926.

Kittler, Friedrich. *Discourse Networks 1800/1900*. Trans. Michael Metteer with Chris Cullens. Stanford, CA: Stanford University Press, 1990.

Knapton, Sarah. "Lying on Facebook Profiles Can Implant False Memories, Experts Warn." *Telegraph*, December 29, 2014. http://www.telegraph.co .uk/news/science/science-news/11315319/Lying-on-Facebook-profiles-can -implant-false-memories-experts-warn.html.

Koepenick, Lutz. *On Slowness: Toward an Aesthetic of the Contemporary*. New York: Columbia University Press, 2014.

Konersmann, Ralf. "Nachwort. Walter Benjamins philosophische Kairologie" [Afterword: Walter Benjamin's philosophical kairology]. In *Kairos. Schriften zur Philosophie*, ed. Ralf Konersmann. Frankfurt: Suhrkamp, 2007.

Koselleck, Reinhart, and Horst Günther. "Geschichte" [History]. In *Geschichtliche Grundbegriffe. Historisches Lexikon zur politisch-sozialen Sprache in Deutschland* [Basic historical concepts: Historical lexicon of political and social language in Germany], ed. Otto Brunner, Werner Conze, and Reinhart Koselleck, 2:593–717. Stuttgart: Klett-Cotta, 1975.

Kracauer, Siegfried. *Mass Ornament: Weimar Essays*. Trans. Thomas Y. Levin. Cambridge, MA: Harvard University Press, 2005.

——. "Photography." Trans. Thomas Y. Levin. *Critical Inquiry* 19, no. 3 (Spring 1993): 421–36.

——. *The Salaried Masses: Duty and Distraction in Weimar Germany*. Trans. Quintin Hoar. London: Verso, 1998.

Kracke, Bernd, and Marc Ries, eds. *Expanded Narration. Das Neue Erzählen* [Expanded narration: The new narration]. Bielefeld: transcript, 2013.

Kristeva, Julia. *Strangers to Ourselves*. Trans. Leon S. Roudiez. New York: Columbia University Press, 1991.

Kross, Ethan, et al. "Facebook Use Predicts Declines in Subjective Well-Being in Young Adults." *PloS ONE 8*, August 14, 2013. http://journals .plos.org/plosone/article?id=10.1371/journal.pone.0069841.

Kucklick, Christoph. *Die granulare Gesellschaft: Wie das Digitale unsere Wirklichkeit auflöst* [Granular society: How the digital dissolves our reality]. Berlin: Ullstein, 2014.

Kuhn, Martin. *Federal Dataveillance: Implications for Constitutional Privacy Protections*. New York: LFB Scholarly Publications, 2007.

Kümmel, Albert. "Innere Stimmen. Die deutsche Radiodebatte" [Inner voices: The German radio debate]. In *Einführung in die Geschichte der Medien* [Introduction to media history], ed. Albert Kümmel, Leander Scholz, and Eckhard Schuhmacher, 175–97. Padeborn: UTB, 2004.

Laclau, Ernesto, and Chantal Mouffe. *Hegemony and Socialist Strategy: Towards a Radical Democratic Politics*. London: Verso, 1985.

Lafargue, Paul. *The Right to Be Lazy and Other Studies*. Trans. Charles Kerr. London: Charles Kerr & Co., 1883.

Landow, George P. *Hypertext 2.0: The Convergence of Contemporary Critical Theory and Technology*. Baltimore, MD: Parallax, 1997.

Lasch, Christopher. *The Culture of Narcissism: American Life in the Age of Diminishing Expectations*. New York: Norton, 1979.

——. *The Minimal Self: Psychic Survival in Troubled Times*. New York: Norton, 1984.

Latour, Bruno. "Technology Is Society Made Durable." *Sociological Review* 38, no. 1 (1990): 103–31.

Laurel, Brenda. *Computers as Theater*. 2nd ed. Indianapolis, IN: Addison-Wesley Professional, 2013.

Leistert, Oliver, and Theo Röhle, eds. *Generation Facebook: Über das Leben im Social Net* [Generation Facebook: On life in the social net]. Bielefeld: transcript, 2011.

Lejeune, Philippe. "Autobiography and New Communication Tools." In *Identity Technologies: Constructing the Self Online*, ed. Anna Poletti and Julie Rak, 247–58. Madison: University of Wisconsin Press, 2014.

Lessing, Gotthold Ephraim. *The Education of the Human Race*. London: Smith, Elder & Co., 1858.

——. *Lessing's Masonic Dialogues (Ernst und Falk)*. Trans. Abraham Cohen. London: Baskerville, 1927.

——. *Nathan the Wise*. Trans. Adolphus Reich. London: A. W. Bennett, 1860.

Lethen, Helmut. *Cool Conduct: The Culture of Distance in Weimar Germany*. Trans. Don Reneau. Berkeley: University of California Press, 2002.

Leung, Lisa Yuk-ming: "Intimacy for 'Deliberative Democracy'? The Role of 'Friendship' in the Participatory Use of Facebook for Activists in Hong Kong." Paper presented at the eighth annual conference of the Asian Studies Association of Hong Kong, Hong Kong Institute of Education, China, March 2013.

Levy, Daniel, and Nathan Sznaider. *Human Rights and Memory*. Princeton, NJ: Princeton University Press, 2010.

Lévy, Pierre. *Collective Intelligence: Mankind's Emerging World in Cyberspace*. Trans. Robert Bononno. Cambridge, MA: Perseus, 1997.

Lingis, Alphonso. *The Community of Those Who Have Nothing in Common*. Bloomington: Indiana University Press, 1994.

Liu, Alan. "Friending the Past: The Sense of History and Social Computing." *New Literary History* 42, no. 1 (2011): 1–30.

Lohr, Steve. *Data-ism: The Revolution Transforming Decision Making, Consumer Behavior, and Almost Everything Else*. New York: Harper Business, 2015.

Lotz, Christian. "Review of Bernard Stiegler, *The Re-Enchantment of the World: The Value of Spirit Against Industrial Populism*, trans. Trevor Arthur (London: Bloomsbury Academic, 2014)." *Marx & Philosophy: Review of Books 2015* (March 11, 2015). http://marxandphilosophy.org.uk/reviewof books/reviews/2015/1754.

Lovink, Geert. *Networks Without a Cause: A Critique of Social Media*. Malden, MA: Polity, 2012.

Lübbe, Hermann. *Im Zug der Zeit: Verkürzter Aufenthalt in der Gegenwart* [In the course of time: Abbreviated stay in the present]. Berlin: Springer, 1992.

Lyotard, Jean-François. *The Postmodern Condition: A Report on Knowledge*. Trans. Geoff Bennington and Brian Massumi. Minneapolis: University of Minnesota Press, 1984.

Maffesoli, Michel. "Erotic Knowledge." *Secessio* 1, no. 2 (Fall 2012). http:// secessio. com/vol-1-no-2/erotic-knowledge.

——. *The Time of the Tribes: The Decline of Individualism in Mass Society*. Trans. Don Smith. London: Sage, 1996.

Manovich, Lev. *The Language of New Media*. Cambridge, MA: MIT Press, 2001.

Marchart, Oliver. *Post-Foundational Political Thought: Political Difference in Nancy, Lefort, Badiou, and Laclau.* Edinburgh: Edinburgh University Press, 2007.

Mayer, Frederick W. *Narrative Politics: Stories and Collective Action.* Oxford: Oxford University Press, 2014.

McLuhan, Marshall. *Understanding Media: The Extensions of Man.* Cambridge, MA: MIT Press, 1994.

McNeill, Laurie. "There Is No 'I' in Network: Social Networking Sites and Posthuman Auto/Biography." *Biography* 35, no. 1 (Winter 2012): 65–82.

Meretoja, Hanna. "Narrative and Human Experience. Ontology, Epistemology, and Ethics." *New Literary History* 45, no. 1 (Winter 2014): 89–109.

——. *The Narrative Turn in Fiction and Theory: The Crisis and Return of Storytelling from Robbe-Grillet to Tournier.* New York: Palgrave Macmillan, 2014.

Mersch, Dieter. *Ordo ab chao—Order from Noise.* Zurich: Diaphanes, 2013.

Meyrowitz, Joshua. *No Sense of Place: The Impact of Electronic Media on Social Behavior.* New York: Oxford University Press, 1985.

Miller, Nancy K. "The Entangled Self: Genre Bondage in the Age of the Memoir." *PMLA* 122, no. 2 (2007): 537–48.

Miller, Vincent. "New Media, Networking, and Phatic Culture." *Convergence 14* (2008): 387–400.

Miousse, Jean-Sebastien B. "How to Get Control on Facebook and How the Algorithms Work." *Science 2.0,* October 19, 2010. http://www.science 20.com/science_and_music_your_ears/blog/how_get_control_face book_and_how_algorithms_work.

Monahan, Torin. "Surveillance as Cultural Practice." *Sociological Quarterly* 52 (2011): 495–508.

Moorstedt, Michael. "Erscanne dich selbst!" [Scan yourself!]. In *Big Data. Das neue Versprechen der Allwissenheit* [Big data: The new promise of universal knowledge], ed. Heinrich Geiselberger and Tobias Moorstedt, 67–75. Frankfurt: Suhrkamp, 2013.

Moretti, Franco. *Distant Reading.* London: Verso, 2013.

Morozov, Evgeny. *The Net Delusion: The Dark Side of Internet Freedom.* New York: Public Affairs, 2011.

Moyn, Samuel. *The Last Utopia: Human Rights in History.* Cambridge, MA: Harvard University Press, 2010.

Mueller, Milton. "Internet Nation?" 2014. http://www.internetgovernance .org/2014/09/05/internet-nation.

Musil, Robert. *The Man Without Qualities.* Trans. Sophie Wilkins and Burton Pike. New York: Knopf, 1995.

Nancy, Jean-Luc. *After Fukushima: The Equivalence of Catastrophes.* Trans. Charlotte Mandell. New York: Fordham University Press, 2014.

——. *Being Singular Plural.* Trans. Robert Richardson and Anne O'Byrne. Stanford, CA: Stanford University Press, 2000.

——. *The Birth to Presence.* Trans. Brian Holmes et al. Stanford, CA: Stanford University Press, 1993.

——. *Die herausgeforderte Gemeinschaft* [The challenged community]. Trans. (German) Esther von der Osten. Berlin: Diaphanes, 2007.

——. *The Inoperative Community.* Trans. Peter Connor et al. Minneapolis: University of Minnesota Press, 1991.

——. "Of Struction." *Parrhesia* 17 (2013). http://www.parrhesiajournal.org/parrhesia17/parrhesia17_nancy.pdf.

Nietzsche, Friedrich. "On the Use and Abuse of History for Life." In *Untimely Meditations*, trans. R. J. Hollingdale. Cambridge: Cambridge University Press, 1997.

——. *Thus Spake Zarathustra.* Trans. Thomas Common. New York: Modern Library, n.d.

Nora, Pierre. "Between Memory and History: Les Lieux de Mémoire." *Representations* 26 (Spring 1989): 7–24.

——. "Reasons for the Current Upsurge in Memory." *Eurozine*, April 19, 2002. http://www.eurozine.com/reasons-for-the-current-upsurge-in-memory.

Page, Ruth. "Reexamining Narrativity: Small Stories in Status Updates." *Text and Talk* 30, no. 4 (2010): 423–44.

Pariser, Eli. *The Filter Bubble: What the Internet Is Hiding from You.* New York: Penguin, 2011.

Parikka, Jussi. *A Geology of Media.* Minneapolis: University of Minnesota Press, 2015.

Pascal, Blaise. *Opuscules et pensées.* Paris: Hachette, 1897.

Paul, Jean. "Biographische Belustigungen" [Biographical amusements]. In *Werke*, part 2, vol. 4, ed. Norbert Miller. Munich: Hanser, 1963.

——. "Briefe und bevorstehender Lebenslauf" [Letters and forthcoming biography]. In *Werke*, part 2, vol. 4, ed. Norbert Miller. Munich: Hanser, 1963.

——. "Clavis Fichtiana." In *Werke*, part 1, vol. 3, ed. Norbert Miller. Munich: Hanser, 1963.

——. *Levana; Or, the Doctrine of Education.* Trans. A. H. London: George Bell and Sons, 1891.

——. "Die Taschenbibliothek" [The pocket library]. In *Werke*, part 2, vol. 3, ed. Norbert Miller. Munich: Hanser, 1963.

———. "Vorschule der Aesthetik" [Introduction to aesthetics]. In *Werke*, vol. 5, ed. Norbert Miller. Munich: Hanser, 1963.

Peirce, Charles S. *Phänomen und Logik der Zeichen* [Phenomenon and logic of signs]. Frankfurt: Suhrkamp, 1983.

Pfaller, Robert. "The Work of Art That Observes Itself." In *Interpassivity: The Aesthetics of Delegated Enjoyment*, ed. Robert Pfaller, 49–84. Edinburgh: Edinburgh University Press, 2017.

Picard, Max. *The World of Silence*. Trans. Stanley Godman. Chicago: Henry Regnery, 1952.

Pinthus, Kurt. "Masculine Literature." In *The Weimar Republic Sourcebook*, ed. Anton Kaes, Martin Jay, and Edward Dimendberg, 518–20. Berkeley: University of California Press, 1994.

Polkinghorne, Donald E. "Narrative and Self-Concept." *Journal of Narrative and Life History* 1, no. 2/3 (1991): 135–53.

Poster, Mark. "Digital Networks and Citizenship." *PMLA* 117, no. 1 (January 2002): 98–103.

Postman, Neil. *Technopoly: The Surrender of Culture to Technology*. New York: Vintage, 1992.

Pschera, Alexander. *800 Millionen. Eine Apologie der sozialen Medien* [800 million: An apologia for social media]. Berlin: Matthes & Seitz, 2011.

Pütz, Friedrich. "Die richtige Diät des Hörers" [The proper diet for the listener]. In *Medientheorie. 1888–1933. Texte und Kommentare* [Media theory, 1888–1933: Texts and commentaries], ed. Albert Kümmel and Petra Löffler, 273–77. Frankfurt: Suhrkamp, 2002.

Raine, Lee, and Aaron Smith. "Social Networking Sites and Politics." *PEW Reports*, March 12, 2012. http://www.pewinternet.org/~/media/Files/Reports/2012/PIP_SNS_and_politics.pdf.

Ramsay, Stephen. "Toward an Algorithmic Criticism." *Literary and Linguistic Computing* 18, no. 2 (2003): 167–74.

Rancière, Jacques. *Aesthetics and Its Discontents*. Trans. Steven Corcoran. New York: Polity, 2009.

———. "The Misadventures of Critical Thought." In *The Emancipated Spectator*, trans. Gregory Elliott, 25–50. Brooklyn: Verso, 2009.

Readings, Bill. *The University in Ruins*. Cambridge, MA: Harvard University Press, 1996.

Reichert, Ramón. "Einführung" [Introduction]. In *Big Data. Analysen zum digitalen Wandel von Wissen, Macht und Ökonomie* [Big Data: Analyses of the digital transformation of knowledge, power, and economy], ed. Ramón Reichert, 9–31. Bielefeld: transcript, 2014.

———. *Die Macht der Vielen. Über den neuen Kult der digitalen Vernetzung* [The power of the many: On the new cult of digital networking]. Bielefeld: transcript, 2013.

Rheingold, Howard. *The Virtual Community: Surfing the Internet.* New York: Minerva, 1995.

Ricœur, Paul. *Time and Narrative.* Trans. Kathleen McLaughlin and David Pellauer. Chicago: University of Chicago Press, 1984.

Rohbeck, Johannes. *Technologische Urteilskraft—Zu einer Ethik technischen Handelns* [Technological judgment: Toward an ethics of technical action]. Frankfurt: Suhrkamp, 1993.

Rosa, Hartmut. *Acceleration and Alienation: Toward a Critical Theory of Late Modern Temporality.* Aarhus: Aarhus University Press, 2010.

———. *Weltbeziehungen im Zeitalter der Beschleunigung: Umrisse einer neuen Gesellschaftskritik* [Relations to the world in the era of acceleration: Outlines of a new social critique]. Berlin: Suhrkamp, 2012.

Rushkoff, Douglas. *Present Shock: When Everything Happens Now.* New York: Penguin, 2013.

Saint-Exupéry, Antoine de. "Flight to Arras." In *Airman's Odyssey*, trans. Lewis Galantière, 281–437. New York: Reynal & Hitchcock, 1942.

Schechtman, Marya. "The Story of My (Second) Life: Virtual Worlds and Narrative Identity." *Philosophy & Technology* 25, no. 3 (2012): 329–43.

Schiller, Friederich, "Die Bürgschaft" [The hostage] (1798). In *Sämtliche Werke*, 1:352–56. Munich: Hanser, 1980.

Schleiermacher, Friedrich Daniel Ernst. "Versuch einer Theorie des geselligen Betragens" [Attempt at a theory of social behavior]. In *Monologen: Eine Neujahrsgabe* [Monologues: A New Year's gift]. Berlin: Holzinger, 2016.

———. *Werke, Auswahl in vier Bänden* [Works: Selection in four volumes], ed. Otto Braun and Johannes Bauer. Leipzig: F. Eckdart, 1913.

Schmid, Wilhelm. "Fitness? Wellness? Gesundheit als Lebenskunst" [Fitness? Wellness? Health as the art of living]. In *Globalisierung im Alltag* [Everyday globalization], ed. Peter Kemper and Ulrich Sonnenschein, 209–17. Frankfurt: Suhrkamp, 2002.

Schmidt, Eric, and Jared Cohen. *The New Digital Age: Reshaping the Future of People, Nations, and Business.* New York: Knopf, 2013.

Schneider, Manfred. *Transparenztraum* [Transparency dream]. Berlin: Matthes und Seitz, 2013.

Scholz, Trebor, ed. *Digital Labor: The Internet as Playground and Factory.* New York: Routledge, 2013.

Schopenhauer, Arthur. "On Thinking for One's Self." In *Essays of Arthur Schopenhauer*, ed. and trans. T. Bailey Saunders, 321–30. New York, A. L. Burt, 1902.

Schulze, Gerhard. *Die Erlebnisgesellschaft. Kultursoziologie der Gegenwart* [The experience society: Cultural sociology of the present]. Frankfurt: Campus, 1992.

Scruton, Roger. "Hiding Behind the Screen." *New Atlantis*, Summer 2010.

Sennett, Richard. *The Corrosion of Character: The Personal Consequences of Work in the New Capitalism*. New York: Norton, 1998.

——. *The Fall of Public Man*. New York: Knopf, 1977.

Silverman, Jacob. "'Pics or It Didn't Happen'—The Mantra of the Instagram Era: How Sharing Our Every Moment on Social Media Became the New Living." *Guardian*, February 26, 2015. http://www.theguardian .com/news/2015/feb/26/pics-or-it-didnt-happen-mantra-instagram-era -facebook-twitter.

Simanowski, Roberto. *Data Love: The Seduction and Betrayal of Digital Technologies*. New York: Columbia University Press, 2016.

——. *Digital Art and Meaning: Reading Kinetic Poetry, Text Machines, Mapping Art, and Interactive Installations*. Minneapolis: University of Minnesota Press, 2011.

——. *Interfictions. Vom Schreiben im Netz* [Interfictions: On writing on the net]. Frankfurt: Suhrkamp, 2002.

——. "Jean Pauls vergebliche Postmodernität" [Jean Paul's futile postmodernity]. In *Jahrbuch der Jean Paul Gesellschaft* [Yearbook of the Jean Paul Society], 61–73. 2013.

——. *Medien und Bildung* [Media and literacy]. Berlin: Matthes & Seitz, 2018.

——. "System und Witz—Jean Pauls Kosmopolitismus als Effekt des sprachphilosophischen Zweifels" [System and wit: Jean Paul's cosmopolitanism as an effect of linguistic-philosophical doubt]. In *Kulturelle Grenzziehungen im Spiegel der Literaturen: Nationalimus, Regionalismus, Fundamentalismus* [Drawing cultural boundaries in the mirror of literatures: Nationalism, regionalism, fundamentalism], ed. Horst Turk, Brigitte Schultze, and Roberto Simanowski, 170–92. Göttingen: Wallstein, 1999.

Simanowski, Roberto, and Luciana Gattass. "Debates in the Digital Humanities Formerly Known as Humanities Computing." *electronic book review*, March 5, 2017. http://electronicbookreview.com/thread/electro poetics/debated.

Singer, Peter. "Visible Man: Ethics in a World Without Secrets." *Harper's*, August 2011. http://harpers.org/archive/2011/08/visible-man.

Smith, Zadie. "Generation Why." *New York Times Book Review*, November 25, 2010. http://www.nybooks.com/articles/2010/11/25/generation-why.

Sontag, Susan. *Against Interpretation*. New York: Farrar, Straus and Giroux, 1966.

——. *On Photography*. New York: Picador, 1973.

Staples, William G. *The Culture of Surveillance: Discipline and Social Control in the United States*. New York: St. Martin's, 1997.

——. *Everyday Surveillance: Vigilance and Visibility in Postmodern Life*. Lanham, MD: Rowman & Littlefield, 2014.

Startwell, Crispin. *End of Story: Toward an Annihilation of Language and History*. Albany: State University of New York Press, 2000.

Stiegler, Bernard. "Care." In *Telemorphosis: Theory in the Era of Climate Change*, ed. Thomas Cohen, 1:104–20. Ann Arbor, MI: Open Humanities Press, 2012.

——. *For a New Critique of Political Economy*. Trans. Daniel Ross. Cambridge: Polity, 2010.

——. "Memory." In *Critical Terms for Media Studies*. Ed. W. J. T. Mitchell and Mark B. N. Hansen, 64–87. Chicago: University of Chicago Press, 2010.

——. *The Re-Enchantment of the World: The Value of Spirit Against Industrial Populism*. Trans. Trevor Arthur. London: Bloomsbury Academic, 2014.

——. *Taking Care of Youth and the Generations*. Trans. Stephen Barker. Stanford, CA: Stanford University Press, 2010.

Strawson, Galen. "Against Narrativity." In *The Self?*, ed. Galen Strawson, 63–86. Malden, MA: Blackwell, 2005.

——. *Selves: An Essay in Revisionary Metaphysics*. Oxford: Oxford University Press, 2009.

——. *The Evident Connexion: Hume on Personal Identity*. Oxford: Oxford University Press, 2011.

Sunden, Jenny. *Material Virtualities: Approaching Online Textual Embodiment*. New York: Peter Lang, 2003.

Sunstein, Cass R. *Republic.com 2.0. Revenge of the Blogs*. Princeton, NJ: Princeton University Press, 2007.

Swan, Melanie. "'Health 2050,' The Realization of Personalized Medicine Through Crowdsourcing, the Quantified Self, and the Participatory Biocitizen." *Journal of Personalized Medicine* 2 (2012): 93–118.

Taylor, Charles. "The Politics of Recognition." In *Multiculturalism: Examining the Politics of Recognition*, ed. Charles Taylor and Amy Gutmann, 25–73. Princeton, NJ: Princeton University Press, 1994.

Thomä, Dieter. *Erzähle dich selbst. Lebensgeschichte als philosophisches Problem* [Narrate yourself: Life history as a philosophical problem]. Munich: C. H. Beck, 1998.

Tieck, Ludwig. *William Lovell*. Trans. Douglas Robertson. 2009. https://docs.google.com/viewer?a=v&pid=sites&srcid=ZGVmYXVsdGRvbWFpbnxoaGV3cmXdhbm5sleHxneDo2MzRiYTFmOTNhMTYrNTlk.

Tiqqun. *Theory of Bloom*. Trans. Robert Hurley. London: LBC, 2012.

Türcke, Christoph. "Die geheime Kraft des Rings" [The secret power of the ring]. In *Lessing. Nachruf auf einen Aufklärer. Sein Bild in der Presse der Jahre 1781, 1881 und 1981* [Lessing: Obituary of an Enlightenment man: His image in the press in years 1781, 1881, and 1981], ed. Klaus Bohnen, 155–62. Munich: Wilhelm Fink, 1982.

Turkle, Sherry. "Identität in virtueller Realität. Multi User Dungeons als Identity Workshops" [Identity in virtual reality: Multiuser dungeons as identity workshops]. In *Kursbuch Internet. Anschlüsse an Wirtschaft und Politik, Wissenschaft und Kultur* [Kursbuch new media: Trends in economy and politics, science and culture], ed. Stefan Bollmann and Christiane Heibach, 315–31. Mannheim: Bollmann, 1996.

——. *Alone Together: Why We Expect More from Technology and Less from Each Other*. New York: Basic Books, 2011.

Twenge, Jean M. *Generation Me: Why Today's Young Americans Are More Confident, Assertive, Entitled—and More Miserable Than Ever Before*. New York: Free Press, 2006.

van Dijck, José. *Mediated Memories in the Digital Age*. Stanford, CA: Stanford University Press, 2007.

——. "'You Have One Identity': Performing the Self on Facebook and LinkedIn." *Media, Culture & Society* 35, no. 2 (2013): 199–215.

van Ham, Peter. "Europe's Postmodern Identity: A Critical Appraisal." In *Global Society in Transition: An International Politics Reader*, ed. Daniel N. Nelson and Laura J. Neack, 189–216. New York: Kluwer Law International, 2002.

Vattimo, Gianni. *Beyond Interpretation: The Meaning of Hermeneutics for Philosophy*. Trans. David Webb. Stanford, CA: Stanford University Press, 1997.

——. *The Transparent Society*. Trans. David Webb. Baltimore, MD: Johns Hopkins University Press, 1992.

Vetere, Frank, Steve Howard, and Martin R. Gibbs. "Phatic Technologies: Sustaining Sociability Through Ubiquitous Computing." In *Proceedings of the CHI-Conference 2005*. http://www.vs.inf.ethz.ch/events/ubisoc2005/UbiSoc%202005%20submissions/12-Vetere-Frank.pdf.

Wanga, Victoria, John V. Tuckera, and Kevin Haines. "Phatic Technologies in Modern Society." *Technology in Society* 33, no. 1 (2012): 84–93.

Walker-Rettberg, Jill. *Seeing Ourselves Through Technology: How We Use Selfies, Blogs, and Wearable Devices to See and Shape Ourselves*. New York: Palgrave Pivot, 2014.

Weinrich, Harald. *Gibt es eine Kunst des Vergessens?* [Is there an art of forgetting?]. Basel: Schwabe, 1996.

White, Hayden. *The Content of Form: Narrative Discourse and Historical Representation*. Baltimore, MD: Johns Hopkins University Press, 1987.

Whitson, Jennifer R. "Gaming the Quantified Self." *Surveillance and Society* 11, no. 1/2 (2013): 163–76.

Wilms, Wilfried. "The Universalist Spirit of Conflict: Lessing's Political Enlightenment." *Monatshefte* 94, no. 3 (2002): 306–21.

Wittkower, Dylan E. "Boredom on Facebook." In *Unlike Us Reader: Social Media Monopolies and Their Alternatives*, ed. Geert Lovink and Miriam Rasch, 180–88. Amsterdam: Institute of Network Cultures, 2013.

Wolf, Gary. "Know Thyself: Tracking Every Facet of Life, from Sleep to Mood to Pain, 24/7/365." *Wired*, July 17, 2009.

Young, Nora. *The Virtual Self: How Our Digital Lives Are Altering the World Around Us*. Toronto: McClelland & Stewart, 2012.

Zhao, Shanyang, Sherry Grassmuck, and Jason Martin. "Identity Construction on Facebook: Digital Empowerment in Anchored Relationships." *Computers in Human Behavior* 24, no. 5 (2008): 1816–36.

Zielinski, Siegfried. *[. . . After the Media]: News from the Slow-Fading Twentieth Century*. Trans. Gloria Custance. Minneapolis, MN: Univocal, 2013.

Zuckerberg, Mark, and Priscilla Chan. "A Letter to Our Daughter." December 1, 2015. http://www.facebook.com/notes/mark-zuckerberg/a-letter-to-our-daughter/10153375081581634.

Zuckerman, Ethan. *Rewire: Digital Cosmopolitans in the Age of Connection*. New York: Norton, 2013.

Zwigenberg, Ran: *Hiroshima: The Origins of Global Memory Culture*. Cambridge: Cambridge University Press, 2014.

INDEX

activism, 15, 143, 146, 220n84
Addresses to the German Nation
(Fichte), 122
Adorno, Theodor W., 18, 19–20, 155,
175n28
advertising, 10, 16, 20, 40, 56, 140,
176n30
aesthetics, 56, 80, 148, 222n94
affective capitalism, 202n53
affective-computing, xv, 169n8
Agamben, Georgio, xiv, 22, 24, 49,
60, 64, 169n7, 176n3, 214n61
agency, 200n50, 201n51, 202n53
aggression, nationalism and, 135
Alexa, 157
Alexander, Matthias, 221n86
algorithm, xvi, 2, 8, 16–18, 70–71,
78, 83, 88–89, 153–55, 171n12,
193n24, 200n50, 201n51, 225n3
alienation, 37, 40, 218n78
Alone Together (Turkle), 137
ambient attention, ix
ambient awareness, 217n75
ambient intimacy, 3
Anderson, Benedict, 39
Annals of St. Gall, 57, 85

Anonymous, 143, 220n84
antagonism, 221n88, 222n94
anxiety, 31, 81
apathy, 27, 48
apps, 24, 73–74, 78, 82, 89, 113,
195n34, 220n84
Arcades Project (Benjamin), 47
archiving: democratization of, 117;
email, 209n35; experts and, 117;
of Facebook, 64; formats,
114–15; history and, 113; internet
and, 111, 114; mania and, 112;
memory and, 112, 209n35;
narrative and, 110; photography
and, 110–11; recording and,
209n35; self, 113–14; storytelling
and, 110; technology, 209n35;
websites and, 113
Arrington, Michael, 175n29
art, community-based, 222n94
artificial intelligence, 76, 155, 156
Assmann, Aleida, 170n7
attentional economy, 11, 26, 61, 65,
168n5
attention deficit disorder, 95
authenticity, 61, 87, 93